Anonymous

Bleaching, Dyeing, and Calico-printing

Anonymous

Bleaching, Dyeing, and Calico-printing

ISBN/EAN: 9783337163952

Printed in Europe, USA, Canada, Australia, Japan

Cover: Foto ©berggeist007 / pixelio.de

More available books at **www.hansebooks.com**

BLEACHING DYEING, AND CALICO-PRINTING.

With Formulæ.

LONDON:
J. & A. CHURCHILL,
11, NEW BURLINGTON STREET.
1884.

EDITOR'S PREFACE.

ALTHOUGH the present volume claims to be little more than a summary of the arts of Bleaching, Dyeing, and Calico-printing, the Editor hopes it may, notwithstanding, be found a ready and serviceable manual for practical workers, and that in many cases, it may be instrumental in shortening the time and trouble, that would otherwise be expended in consulting the pages of larger and more elaborate works.

The Editor has to tender his grateful acknowledgments to his friends, Mr. J. F. HODGES, jun., of Belfast, and Mr. JAMES CHADWICK, of Manchester; the former for his valuable assistance in the compilation of the chapter on "Bleaching," the latter for his no less valuable contributions to the sections on "Dyeing and Calico-printing." To Mr. CHADWICK the Editor is indebted for many of his formulæ, and additionally for a large amount of equally useful technical information.

Lastly, the Editor has to acknowledge his indebtedness to Mr. CROOKES' "Practical Handbook of Dyeing and Calico-printing;" URE's "Dictionary of Arts, Manufactures, and Mines;" WAGNER's "Chemical Technology," edited by W. CROOKES, F.R.S.; CALVERT's "Dyeing and Calico-printing," edited by STENHOUSE and GROVES, F.R.S., and SPON's "Encyclopædia of the Industrial Arts:" all of which works have been consulted in the preparation of this little volume.

JOHN GARDNER, F.I.C., F.C.S.

LONDON, *December*, 1883.

CONTENTS.

CHAP.		PAGE
I.	Bleaching	1
II.	Dyeing	33
III.	Calico-Printing	61
IV.	Dye Stuffs	133
	Appendix	193

BLEACHING, DYEING,

AND

CALICO PRINTING.

CHAPTER I.

BLEACHING.

BLEACHING is the process by which the colour of bodies, natural or acquired, is removed, and by which they are rendered white or colourless. It is more particularly applied to the decolorization of textile filaments, and of cloths made of them.

Bleaching is a very ancient art, as passages referring to it in the earlier sacred and other writers fully testify. It had probably reached a high degree of excellence among the inhabitants of the first Assyrian empire, and was certainly practised in Egypt long before the commencement of written history. We may fairly assume that fine white linen formed part of the "raiment," which, together with "jewels of gold and jewels of silver and precious things," Abraham sent as presents to the beautiful Rebekah and her family, fully three centuries and a half before the Exodus. Subsequently, in Scripture, we have special mention of "fine linen, white and clean." Herodotus, the earliest Greek historian, tells us that the Babylonians wore "white cloaks;" and in Athenæus we read of "shining fine linen," as opposed to that which was "raw" or unbleached. At this early period, and for many centuries afterwards, the operations of washing, fulling, and bleaching were not distinctly separated. The common system of washing, followed by drying in the sun, adopted by the ancients, is a process which of itself, by frequent repetition, decolorizes the raw materials of textile fabrics, and thus must inevitably have taught them the art of "natural bleaching" of a character similar to that practised in Europe up to a comparatively very recent period. And this appears, according to the authority of ancient authors, to have been the case. Washing or steeping in alkaline and ammoniacal lyes, or in milk of lime, followed by exposure in the sun, formed the chief basis of their system; whilst woollens, then as now, were treated with soap and fuller's earth, or with potter's clay, marl, limolian earth, or other like minerals. Urine was highly esteemed among them; and we are told that, in the time of the Emperor Vespasian, and undoubtedly long before, cloths were sulphured. Indeed, according to Pliny, sulphuring was often had recourse to in ordinary washing, as well as in the bleaching process. Bleaching continued to be practised with no essential change of its principles until the discovery of chlorine, to which we shall presently refer.

Though the art of bleaching dates from the remotest ages, little or nothing was known of it in Great Britain as an art, until within a century, and it was then the custom to send the linen manufactured in Great Britian over to Holland, as the inhabitants of that country were at this period considered the most superior bleachers in Europe. An account of the method which the Dutch adopted at their great bleach works near Haarlem, is described by Mr. F. Hodges, jun., in his "Chemistry for Bleachers." It consisted in steeping the goods in a lixivium or lye made from the lye-ashes of Russia, and in which other cloth had been steeped; afterwards they were steeped in a new lye of lye-ashes poured upon them boiling hot. In this solution they were left for some days, after which they were washed and pressed, and then steeped in a sour made either by the fermentation of bran and water or buttermilk. The souring usually lasted from six to seven days; after which the linen was washed, and then spread upon the grass to bleach by exposure to light, air, and moisture. The bleaching grounds were cut with canals in different places, from which the linen was watered with long narrow shovels made in the shape of a scythe; the water of these canals came from the sand downs, and to the beneficial effects derived from it was mainly attributed the superior lustre of the Dutch cloth; indeed, it was long a prejudice on the Continent that no water was so good for bleaching as sea water. This process usually required from six to seven months for its completion, and the goods bleached by it were sold under the name of Dutch cloth or Hollands.

Another variety of linen bleached at Haarlem, which from its fineness was generally spread out on the better grass fields or lawns, received the title of lawn. Several authorities relate that in 1749 an Irishman named DUNLOP, who had learned something of the nature and art of bleaching, settled in the north of Scotland, and established works for the purpose of bleaching Scotch goods; and though for some years he failed to bleach the goods entrusted to him satisfactorily, in a few years he became an excellent practical bleacher, and from that time no more goods were sent to Holland for the purpose of being bleached. The art, when introduced into Great Britain by this enterprising Irishman, is described as not differing in the smallest degree from the method employed by the Dutch, from whom it was copied. It consisted of steepings and boilings in alkaline lyes, called bucking, then washing and exposing on grass, called crofting; these operations were repeated several times, reducing the strength of the lyes every time. The linen was then steeped in sour milk for some weeks, after which it was washed clean and crofted. This process was repeated as often as was required to produce a pure white. The first improvement in this tedious process occurred after the introduction of a new method for the manufacture of sulphuric acid by Dr. ROEBUCK, which greatly reduced the price of that acid. The improvement consisted in the substitution by Dr. FRANCIS HOME, of Edinburgh, of water acidulated with sulphuric acid, as a sour for the buttermilk hitherto employed; this reduced the time required for souring from weeks to days, not to mention the absence of the risk with which the use of the milk was accompanied, as sours of this description were very liable to corruption. Little further change took place in the art of bleaching until about the year 1787, when a most important improvement was effected in consequence of the discovery, in the year 1774, by SCHEELE, the celebrated Swedish chemist, of a substance which he called dephlogisticated marine air. The French chemist, BERTHOLLET, in 1785, repeated the experiments of SCHEELE on this new substance, and showed that it was a gas soluble in water, to which it gave a yellowish green colour, an astringent taste, and the peculiar smell by which it is distinguished. Owing to BERTHOLLET's experiments, this body was known until the year 1810 as oxygenated muriatic acid, or oxy-muriatic acid, into which it was shortened; but in that year Sir H. DAVY, adopting the idea of GAY-LUSSAC and THÉNARD, who had experimented with it, that it was an elementary substance, called it chlorine, owing to its peculiar colour, by which name it is known at the present day. BERTHOLLET's ideas, and the result of some experiments which he made in bleaching linen

with the gas, were mentioned by him in a paper which he read
before the Academy of Sciences, at Paris, in April, 1795, and
published in the *Journal de Physique* for May of the same
year. He also published a paper in the number for August,
1786, of the same journal, explaining the nature of the action
of chlorine on vegetable colours, and showing how it could best
be employed. PARKES, in his " Chemical Essays," relates how
a Mr. COPLAND, Professor of Natural Philosophy in Aberdeen,
while on a visit to Geneva, was shown by Professor de
SAUSSURE of that town, the experiment of discharging
vegetable colours by chlorine gas. The Aberdeen Professor,
having been impressed with the importance of the experiment, communicated it on his return home to some eminent
manufacturers, the Messrs. MILNES, of the firm of GORDON,
BARRON & Co., of Aberdeen, who immediately entered upon
a course of experiments in the preparation of the gas, and
the best manner of employing it in bleaching, and obtained
satisfactory results. PARKES states that this was about the
end of July, 1787, and was, he believes, the first actual
application of the then so-called oxy-muriatic acid in Great
Britain. With this statement, however, other authorities
on the subject do not agree, and it seems with some truth.
Mr. F. HODGES, jun., in his "Chemistry for Bleachers,"
goes fully into the whole subject, and the weight of evidence
which he brings to bear from numerous quarters goes far to
prove that the distinguished engineer, JAMES WATT, if not
the first, at least is entitled to the honour of having introduced it at as early a date as Professor COPLAND. WATT
learnt the process of manufacturing and using chlorine from
BERTHOLLET in 1786, and shortly afterwards introduced the
process on a large scale into his father-in-law's (Mr. MAC
GREGOR) bleach works at Glasgow. WATT laid the results of
the employment of this gas in bleaching at Glasgow, before
the Manchester manufacturers. In enforcing the importance

of the new substance and process on these gentlemen, he was ably followed and seconded by Mr. THOMAS HENRY, F.R.S., of Manchester, and it is related how this gentleman and WATT unreservedly described to each other the result of their experiments. To Mr. HENRY we are indebted for the introduction of the new art into Lancashire. WATT made several improvements in the art, one of which was that instead of employing muriatic acid and manganese, as had been done by SCHEELE and BERTHOLLET for the production of the gas, he used a mixture of common salt, black oxide of manganese and sulphuric acid, which was much cheaper. He also invented a method of testing the strength of the water impregnated with chlorine, so as to estimate its bleaching power. This he did by taking a known quantity of infusion of cochineal, and ascertaining how much of the bleaching liquor was necessary to destroy the colour—the larger the quantity necessary, the weaker obviously was the bleaching solution. Chlorine, when first introduced, was used in the state of gas, and one great drawback to it was its noxious odour, which is not only very disagreeable, but exceedingly injurious to health.

To BERTHOLLET we owe the credit of being the first to remedy many of the defects of bleaching by chlorine, as, while visiting Javelle for the purpose of showing some bleachers the method of using the gas, he added a little potash to prevent the gas from impairing the goods. Not long after this, these bleachers announced in different journals that they had discovered a new bleaching liquor, which they called the "Lye" or "Eau de Javelle," and applied to the British Government to grant them the exclusive right to supply the public with it, but in this they were defeated, as it was shown that the same article had been in common use in Great Britain for some time, which fact prevented them from obtaining a patent, and consequently the liquor of Javelle, which thus became the property of the public, turned out to be nothing but a solution of potash in water impregnated with chlorine, as was proved by BERTHOLLET shortly after its pretended discovery. After the failure of the foreigners to obtain a monopoly of the Lye of Javelle, other bleachers learnt to make it for themselves, and continued to use it for some time. Though this bleaching liquor had some advantages over the solution of the gas in water, they were more than counterbalanced by the disadvantages, such as its being less economical than the solution of chlorine in water, and its not keeping any length of time without losing its bleaching properties. On account of these disadvantages, its use was not long continued.

The next attempt to improve on this bleaching solution was made by Mr. HENRY, of Manchester, to whom we have before referred, who is said to have first thought of the addition of lime, but owing to his manner of employing it, which was open to many objections, it did not come into use. Other attempts were made by different persons to improve on this process, but none succeeded till Mr. TENNANT, after long and laborious investigation, hit upon a method of making a saturated liquid, composed of chlorine and lime, for which he took out a patent in the year 1798. This patent was pronounced invalid, and unjustly so, as many authorities consider. But Mr. TENNANT was not so easily defeated, for in the following year, 1799, he took out another patent, which may be considered the completion of the new method. This patent consisted in impregnating quicklime in a dry state with chlorine. As the originality of this invention was not disputed, and as its great superiority over all methods previously introduced was obvious, the demand for the product has gone on increasing, year by year, up to the present date. The "new or continuous process" of bleaching, as it is called, and that which is at present in general use in all the chief bleach works of Lancashire, was introduced by Mr. DAVID BENTLEY, of Pendleton, and patented by him in 1828.

Bleaching is commonly said to be natural when exposure to light, air, and moisture forms the leading part of the process; and to be chemical when chlorine, or any of the hypochlorites, or sulphurous acid, or other like substances, are employed. In some cases, as with linen, the two processes are combined.

The subject will be noticed under separate heads, depending on the material operated on.

I. **Bleaching of Cotton.**—Cotton is more easily bleached, and appears to suffer less from the process, than most other textile substances. On the old plan it was first thoroughly washed in warm water to remove the weaver's paste or dressing; then bucked or "bowked" (boiled) in a weak alkaline lye, or in milk of lime, to remove colouring, fatty, and resinous matters insoluble in simple water; and after being again well washed, was spread out upon the grass, or bleaching ground, and freely exposed to the joint action of light, air, and moisture (technically called "crofting"). The operation of "bucking" in an alkaline lye, washing, and exposure was repeated as often as necessary, when the goods were "soured" or immersed in water acidulated with sulphuric acid, after which they received a final thorough washing in clean water, and were dried, finished, and folded for the market. From the length of the exposure upon the bleaching-ground, this method is apt to injure the texture of the cloth; and from the number of operations required is necessarily expensive and tedious. It is, therefore, now very generally superseded by the system of chemical bleaching, briefly described below. In the chemical system of bleaching, the goods are washed and "bucked" as on the old plan, then submitted to the action of a weak solution of chloride of lime, and afterwards passed through water soured with hydrochloric or sulphuric acid, and again thoroughly washed, then dried and finished. The aim of the old as well as the new process is to remove with the smallest amount of risk to the goods and at the lowest cost, as well as in the shortest time, the natural as well as the acquired impurities of the cotton. The nature of the former, which amount to only about 1 per cent., has been carefully studied by Dr. SCHUNCK, and they have been found by him to consist of fatty and waxy matters, brownish colouring substances, pectic acid, and albuminous matter. The acquired impurities consist of all those foreign matters obtained either accidentally or intentionally during the process of manufacture. Their amount varies enormously, often exceeding 30 per cent. They consist of the various matters, organic and inorganic, introduced during the sizing of the warps, such as china clay, magnesium chloride, zinc chloride, starch or flour, grease from the size, the machinery, and the hands of the workmen, and dust and dirt of all kinds. The new or continuous process, before referred to, is the method of chemical bleaching at present in most general use; and, indeed, it has nearly superseded all other methods. In this system, the pieces previously tacked together endwise, so as to form a chain, are drawn by the motion of rollers in any direction, and any number of times through every solution to the action of which it is desired to expose them; being at the same time entirely and completely under the control of the operator. Annexed is an outline of the several operations in the improved form of the continuous process as practised by Messrs. McNAUGHTON, BARTON & THOM, at Chorley, and in most other large bleach-works.

1. *Preliminary Operations:* — *a.* The "pieces" are separately stamped to enable the bleacher to distinguish

between different lots of cloth, and to detect faults. The goods in the grey state are marked by any colour (usually black), which will sufficiently resist the bleaching process. Among the marking materials which are commonly employed, are gas-tar, pretty thick, alone or mixed with lamp-black; a solution of nitrate of silver, boiled oils coloured with red lead or lamp-black; and aniline black; but this latter has many objections, such as a tendency to produce holes. In linen bleaching, each piece is marked with letters made by red thread, which is found to stand the bleach better than any of the above agents.

b. They are tacked together endwise, either by hand or a machine, so as to form one continuous piece of 300 to 350 yards in length, according to the weight of the cloth.

c. The next operation is singeing, the object of which is to remove all the fine loose down from the surface of the cloth. This is accomplished either by passing the goods rapidly over revolving hot cylinders, by hot plate, by coke flame, or by gas flame. The plan most generally adopted is by the gas machine, invented by TALPIN, and improved upon by Messrs. MATHER & PLATT.

d. 1. After the singeing, the goods are washed; either before or after washing, they are crushed into a rope-like form by drawing them through a smooth aperture, the surface of which is generally of glass or porcelain, the rope-form being given them to enable the water and other liquids to penetrate the goods more easily, and to allow them to be laid in loose coils in the kiers.

2. The following process requires to be modified slightly, if a market bleach or madder-bleach is desired; the latter is the name given by bleachers when the goods are required as white as possible, the former if the goods are to be sold in a white state, and not printed before going into the market. The goods are run through milk of lime, contained in the

"limeing machine," direct into the kier, where they are bucked or boiled under pressure from twelve to fourteen hours, followed by rinsing or cleansing in the washing-machine.

3. They are soured in water acidulated with hydrochloric acid; this process is known as the "lime sour" or "grey sour," after which they are again washed by the washing-machines.

4. They are now bucked or boiled for fifteen or sixteen hours in a solution of resinate of soda, and then washed as before. In some works they give the goods a light boil with soda ash, free from caustic, after the resin boil; this is to remove any risk of resin remaining in the cloth.

5. They are chemicked by being laid in a wooden, stone, or slate cistern, when a solution of chloride of lime is pumped over them, so as to run through the goods into a vessel below, from which it is returned on them by continued pumping, so that the cloth lies in it for one or two hours. This operation requires great care, particularly in the preparation of the chloride of lime solution; as Mr. F. Hodges, jun., has shown that if the smallest particle of undissolved bleaching powder is allowed to come into contact with the cloth it is liable to produce holes. The goods are again washed.

6. They are bucked or boiled for four or five hours in a solution of 1 lb. of crystallized carbonate of soda, dissolved in 5 gallons of water, to every 35 lbs. of cloth, and washed.

7. They are again "chemicked" as before, and washed.

8. They are soured in very dilute hydrochloric acid, and then left on stillages for five or six hours.

9. They are finally thoroughly washed, well squeezed between rollers, dried over steam heated tin cylinders, starched or dressed, and finished. This is the usual process for good calicoes. Muslins, and other light goods, are

handled rather more carefully; whilst for commoner goods the sixth and seventh operations are generally omitted. The entire process usually occupies five days; but by using Mr. BARLOW's high pressure steam-kiers it may be performed in two. PENDLEBURY's kier, which is not unlike BARLOW's, is generally used when working on a small scale. Yarns, as they contain a smaller percentage of artificial impurities than cloth, are bleached in a somewhat different manner. The skeins are first looped together, after which they are boiled in open kiers in soda lye, then in water, washed, chemicked, washed, soured and washed.

According to the most reliable authorities, the strength of cotton fibre is not impaired by its being boiled for two hours in milk of lime, under ordinary pressure, out of contact with the air; nor, according to the bleachers, even by sixteen hours, boiling at the strength of 40 lbs. per 100 gallons. It is said that lime is less injurious than soda. Solution of caustic soda, sp. gr. 1·030, does not injure it, even by boiling under high pressure; but, in practice, soda-ash, or carbonate of soda, is used, and this is only in the second bucking, and in the third, if there be one. The strength now never exceeds 25 lbs. of the crystals to the 100 gallons, and is usually less. Experiments have shown that immersion for eight hours in a solution of chloride of lime,* containing 3 lbs. to the 100 gallons, followed by souring in sulphuric acid of the sp. gr. 1·067,

---

* Since the introduction of bleaching powder many chemicals have been proposed as a substitute for this bleaching agent, but up to the present not one of them has met with any but a partial success. Among those brought before the public may be mentioned permanganate; chlorozone, obtained by passing a mixed current of hypochlorous acid and air through a solution of caustic soda, this compound being considered by some to bleach better than the hypochlorites, the action of which in bleaching it resembles; chlorochromic acid, the chlorates, and peroxide of hydrogen. Perhaps the or for eighteen hours in acid of 1·035, does not injure it. By the improved method of previously treating the goods with lime or alkalies, little chloride of lime is required. Indeed, it is said that where 300 lbs. were formerly employed, 30 to 40 lbs. only are now used. At the same time it is right to mention, that though a solution at $\tfrac{1}{2}°$ Twaddle is usually regarded as the best and safest strength, yet in some bleach works, particularly for inferior and less tender goods, this is greatly increased, even up to 5°, the period of immersion being proportionally reduced, as it is not safe to expose the goods long to the action of such powerful solutions. With the higher strengths they are passed rapidly through the liquid with the callender, sufficient time only being allowed to soak them thoroughly, then immediately through the acid or souring, followed by washing as before.

2. **Bleaching of Linen.**—Linen may be bleached in a similar way to "cotton," but the process is much more troublesome and tedious, owing to its greater affinity for the colouring matter existing in it in the raw state. Under the old system several alternate buckings with pearlash or potash and lengthened exposure on the field, with one or two sourings, and a final scrubbing with a strong lather of soft soap, constituted the chief details of the process. In this way a high degree of whiteness, though not an absolutely pure or snow white, was ultimately produced. Grass-bleaching or crofting is still extensively used for linen; but

---

most novel plan proposed as a substitute for the chemicking, is that invented by ENGLER, who bleaches with the vapours of chloroform, generated by means of a mixture of quicklime, chloride of lime, alcohol or acetic acid, sulphuric acid, and water. In Scotland and Ireland the washing is generally performed by wash-stocks, whilst in Lancashire dash-wheels or washing-machines with squeezers are almost always used for the purpose. Cotton loses about 1-20th of its weigh by bleaching.

it is more generally employed only for a limited time, and in combination with a modification of the system at present almost universally adopted for cotton goods; whilst in some cases crofting is omitted altogether, and the bleaching conducted wholly by the latter process. Linen goods are bleached either in the form of yarn, thread or cloth. The following Tables exhibit the outlines of the new method as at present practised in Ireland and Scotland for linen bleaching. Table No. I. is that which is given as suitable for a parcel of light linens, by a well-known proprietor of an Irish bleach works, at a time when the now but little used fermentation process was in favour. No. II. is the outlines of the new system at present practised in Ireland and Scotland for plain sheetings.

Table No. I.

1. Steep for 24 hours, wash 15 minutes; time 2 days.
2. Boil for 7 hours in lye and resin $2\frac{1}{2}°$, wash 15 minutes; time 2 days.
3. Boil for 9 hours in lye $2\frac{1}{2}°$, wash 30 minutes; time 1 day.
4. Grass for 3 days; time 3 days.
5. Boil for 10 hours in lye $3°$, wash 30 minutes; time 1 day.
6. Grass for 3 days; time 3 days.
7. Boil for 8 hours in lye $3°$, wash 30 minutes; time 1 day.
8. Grass for 3 days; time 3 days.
9. Rough sour for 10 hours in vitriol $2°$, wash 40 minutes; time 1 day.
10. Scald for 4 hours in weak lye, wash 30 minutes; time 1 day.
11. Grass for 2 days; time 2 days.

12. Dip for 10 hours in alkali 40 to 1 strength, wash 30 minutes; time 1 day.
13. Sour for 12 hours in vitriol $1\frac{1}{2}°$, wash 45 minutes; time 1 day.
14. Scald for 4 hours in lye and soap, wash 20 minutes; time 1 day.
15. Rub with brown soap, wash 35 minutes; time 1 day.
16. Grass for 2 days; time 2 days.
17. Dip for 10 hours in alkali 30 to 1 strength, wash 20 minutes; time 1 day.
18. Sour for 12 hours in vitriol 1°, wash 45 minutes; time 1 day.
19. Scald for 3 hours in soap and lye, wash 30 minutes; time 1 day.
20. Dip for 10 hours in alkali 45 to 1 strength, wash 20 minutes; time 1 day.
21. Sour for 12 hours in vitriol 1°, wash 45 minutes; time 1 day.
22. Rub with soap.

Time taken 31 days.

The goods should now be white and ready for finishing.

Table No. II.

a. For plain sheetings :—
1. They are bucked for 12 or 15 hours in a lye made with about 1 lb. of pearlash (or soda-ash) to every 56 lbs. of cloth, and washed.
2. Crofted for about 2 days.
3. Bucked in milk of lime.
4. Turned, and the bucking continued, some fresh lime and water being added, and washed.
5. Soured in dilute sulphuric acid at 2° Twaddle.
6. Bucked with soda-ash for about 10 hours, and washed.
7. Crofted, as before.

8. Bucked again with soda-ash, as before.
   9. Crofted for about 3 days.
   10. Examined, the white ones taken out, and the others again bucked and crofted.
   11. Scalded or simmered in a lye of soda-ash of about only 2-3rds the former strength, and washed.
   12. Chemicked, for 2 hours, at $\frac{1}{2}°$ Twaddle, washed, and scalded.
   13. Again chemicked, as before.
   14. Soured for 4 hours, as in No. 5; washed, and finished.

   This occupies 13 to 15 days, according to the weather.

b. For shirtings, &c.:—As the preceding, but with somewhat weaker solutions.

c. For goods to be subsequently printed:—
   1. Bucked in milk of lime for 10 or 12 hours.
   2. Soured in dilute hydrochloric acid of 2° Twaddle, for 3 to 5 hours, and washed.
   3. Bucked with resinate of soda for about 12 hours.
   4. Goods turned, reboiled as before, and washed.
   5. Chemicked at $\frac{1}{2}°$ Twaddle, for 4 hours.
   6. Soured at 2° Twaddle, for 2 hours, and washed.
   7. Bucked with soda-ash for about 10 hours, and washed.
   8. Chemicked as in No. 5.
   9. Soured, as at No. 6, for 3 hours; washed, and dried.*

The chief difficulty in bleaching linen arises from the fact that it contains a much larger proportion of natural impurities than cotton. To thoroughly understand the nature of these impurities it would be necessary to study the various processes through which the flax plant passes before being manufactured into linen, this our space will

---

* The strengths of the solutions, when not otherwise stated, are about the same as those given under Cotton.

not permit; it will be sufficient to mention that after the retting process, which Professor HODGES has shown removes from the fibre upwards of 41·1 per cent. of the nitrogenized and other constituents of the plant, there are to be seen numerous brilliant scales of a resinous appearance and light amber hue which are deepened in colour by alkalies, in which also they can be entirely dissolved. The nature of these scales is not as yet clearly understood. The constituents of dressed and undressed flax have been carefully studied by HODGES, sen., and HODGES, jun.; to the investigations of the former we are principally indebted for information as to the nature of these bodies. HODGES, sen., upwards of thirty years ago, gave not only analyses of the gases evolved in the steeping process, but investigated the nature of the constituents of dressed flax, showing that the latter consisted of wax, volatile oil and acid resinous matter, sugar and colouring matters, gum, pectin, nitrogenized compounds, inorganic matters and cellular fibre. HODGES, jun., has lately thoroughly investigated the nature and chemical constitution of these bodies, and finds that the wax is a complex body closely resembling in composition palmitic ceryl ether. To remove these impurities, and to produce a pure white cellulose, is the object of all the bleaching processes described; the machinery which is used to accomplish this in linen bleaching differs greatly from that used by the cotton bleacher. Space will not permit us to enter fully into a description, but the following sketch may be serviceable to those unacquainted with this part of the subject.

WASH MILLS.—These machines which are used in Ireland for washing linen are there called wash mill feet or stocks. They are made by suspending from a strong frame two large blocks of wood, weighing about 534 lbs. each; by means of the water supply, these blocks are made to rise and fall alternately, in this way the goods are first soaked with water and then squeezed free from it. Though the appearance of these machines is far from ornamental, yet they have been found to work more satisfactorily than any other plan. The dash-wheel is a cylindrical box revolving upon its axis, and divided into four compartments, these have openings into which two or more pieces are introduced; water is admitted through the hollow axis and the wheel is set in motion, causing the goods to pass from side to side. Many other washing machines have been proposed, but as yet have met with only very partial success.

RUBBING MACHINE.—This machine is a special feature in linen bleaching. The object of the process called "rubbing" is to give the goods a thorough soaping, to neutralize any injurious acids or alkalies they may contain, and to remove the yellowish brown specks known as sprits. At the foot of this machine there is a trough containing a strong solution of soap kept constantly boiling by steam, the goods are first passed through this trough, then between two square flat pieces of wood or marble, toothed transversely. The upper piece is made to move lengthwise over the goods, which are pulled through by the drawing engine. A description of the yarn and linen drying and finishing machinery would be altogether out of place in a notice such as this, and we shall have to content ourselves with a short account of the several chemical processes described, such as the Steep, the Boil, the Sour, the Scald, and the Dip. Linen goods contain about 33 per cent. of impurities, consisting of those before mentioned; together with these it contains a dressing employed by the weaver, this dressing is used to make the linen smooth and cover bad work, and is made usually from flour in the form of paste; but just as the cotton bleacher has to contend with a numerous list of articles used as a size or as dressing, so has the linen bleacher; among those used by the linen weaver may be mentioned mashed potatoes allowed to sour, a mixture of glue, sago, and tallow boiled together, Irish moss, and soap. These are either used alone or mixed with numerous substances, such as chloride of calcium or magnesium, gum, glycerine, wax, &c. The object of the steep or boil is to remove all these matters, and the success of the bleach in great part depends upon whether this has been successfully accomplished.

The now little used steeping process was carried out in a kieve or box made of wood, stone, or cement, containing water at a temperature of 120° to 150° Fahr. (49° to 65·5° C.), in which the goods were immersed from one to two days, and fermentation was promoted either by pipe clay, resin, bran, infusion of malt, or yeast. When the goods had been sufficiently long in the steep they were washed, and then boiled under pressure, first in old soda lye, then again washed and again boiled in lye cleaner than that at first used, yet not quite pure. The number of boils and their length depended not only on the class of goods to be bleached, but upon the ideas of the bleacher. Lime is now used in most of the Irish bleachgreens for boiling, and though some of the older bleachers do not care for it, yet for many classes of goods it has been found to produce a better result than soda lye for the first boils. It is most important that the goods after boiling in soda lye be not allowed to lie exposed to the air too long without being washed, as the carbonate of soda is apt to crystallize within the fibres, which are burst during the formation of the crystals. The loss of weight by the boiling in lime and caustic or carbonated alkalies, varies according to the class of yarn and its former treatment, from 14 to 37 per cent. The larger the number of boils the greater will be the quantity of the brownish colouring matter of the fibre which will be removed, and after a time the goods will only retain a light grey shade which is without difficulty removed by steeping in a weak solution of a hypochlorite; care must, however, be taken that the hypochlorite be not used until the brownish colour has been removed, as it will be "set" or fastened by the hypochlorite. Steeping in a solution of a hypochlorite is technically called the DIP. The hypochlorite principally used in Ireland for the dip, is for linen, hypochlorite of soda, commonly called chloride of soda; this the bleacher usually makes for himself by adding to a clear solution of bleaching powder or hypochlorite of lime, a solution of soda ash or carbonate of soda, allowing the mixture to settle and drawing off the clear solution, which is the chloride of soda. For yarns and threads the bleaching agents used are hypochlorites of lime or magnesia.

THE SOUR.—Souring consists in immersing the goods from four to eight hours in a bath of dilute sulphuric acid and then washing. It is called a rough sour if the goods are soured after coming from the grass and before they have had a dip.

THE SCALD.— Scalding is a light boil of from two to three hours in a clean and weak lye containing some soap, together with the soap left in the goods after the rubbing process. The above is an outline of the process and machinery at present employed in Ireland for linen bleaching, and with the exception of the machinery, which has been greatly improved, little or no change has taken place in the method since the introduction of chlorine. Though many plans have been proposed to shorten the process by doing away with the grassing, none so far as linen is concerned have succeeded; the latest and perhaps the most novel plan suggested is that lately patented by Messrs. J. J. DOBBIE & J. HUTCHESON, who generate chlorine by the electrolysis of dilute hydrochloric acid. Their process consists in steeping the cloth to be bleached in sea-water, and passing the fabric between a series of carbon rollers, the upper row of which is connected with one pole, the lower row with the other pole of a battery. The rollers are caused to rotate slowly, and thus pass the fabric from one end to the other. Hypochlorite is formed, and on subsequent immersion in acid the fabric is effectually bleached.

For yarn and thread bleaching the process which has been found most successful is that of HODGES, jun., which is known in Ireland as the "Chemico-Mechanical Process," so called from the patentee turning to account the advantages derivable from the employment of mechanical contrivances driven by steam, combined with the introduction of a new method of obtaining the hitherto little used hypochlorite of magnesia. This process may be said to date from the discovery of the substance known as *Kieserite* (native sulphate of magnesia), which occurs as an essential constituent of the Abraum salts of Stassfurt. For some time after the introduction of this substance into the market, it was considered of little value except for the production of Epsom salts; but Mr. HODGES, in the course of some investigations in bleaching jute, having had occasion to employ large quantities of hypochlorite of magnesia, it occurred to him that kieserite might be substituted for the more expensive crude sulphate of magnesia; and the importation into Ireland of the sample for this purpose, was the first that was ever sent into that country for the manufacture of a bleaching liquor, or, indeed, for any other use. Mr. HODGES, on experimenting with the kieserite, found that it not only supplied the place of the crude sulphate, but acted as a better precipitant for the lime of the bleaching powder, which is employed in the production of the hypochlorite of magnesia; and that it also produced a stronger and clearer solution. Without entering into a minute description of the process, the following outline will be sufficient to show the nature of the methods adopted. The kieserite, which is imported from

Germany in square blocks, on arriving at the works, is conveyed to a house, on the ground-floor of which it is stacked until required, when it is ground to a fine powder, placed in barrels, and drawn up by means of a crane to a room at the top of the building, at one end of which is a row of three tanks furnished with water taps, agitators, and false bottoms. In one of the end tanks a definite quantity of the kieserite powder (varying according to its strength, ascertained by analysis) is placed and dissolved in a given quantity of water, the solution being assisted by agitators, and on settling, the clear liquor is siphoned over into the middle tank. In the third tank, bleaching powder (hypochlorite of lime), varying in quantity according to the strength of the kieserite solution, is placed. The bleaching powder after being agitated with water is allowed to settle, and the clear solution is siphoned over into the middle tank containing the clear kieserite solution, the agitator being kept in motion, not only during the mixing of the liquids, but for some time after. The mixed liquids are then allowed to remain undisturbed all night, after which the clear hypochlorite of magnesia solution is siphoned into a large settling tank, which is situated in the room below. From this vessel it is conducted through wooden pipes (which are so contrived that they can be opened and cleansed at will), into a large cistern standing in the bleaching-house. This cistern is fitted with a ball-cock, by which arrangement the liquid can be drawn off by a system of wooden pipes as required. The bleaching-house in which the cistern is situated is fitted up in an original manner, and covers something more than an acre of ground; whilst the reeling-shed, which is the only part of the works our limits will permit us to describe, is 240 feet long by 24 feet broad, and contains ten steeps and twelve reel boxes. Each box is provided with water, a solution of the bleaching agent, and steam pipes, and is capable of reeling at a time about 500 lbs. of yarn. Above the box is a line of rails on pillars. A travelling crane runs along the rails, and carries the reels from one box to another. Attached to this crane is a newly invented hydraulic pump, by means of which the reels with the yarn on them can be lifted in a few seconds from one box to another.

After the yarn has been boiled, washed, and passed through the squeezers in the usual manner, it is put on a waggon, in which it is carried, by means of a line of rails, down to the first reel box. Here it is placed on the reels, which are made to revolve by means of steam first in one direction and then in another, through a solution of carbonate of soda, previously heated by means of the steam-pipes before mentioned. The yarn having been sufficiently scalded and so saturated with soda, the reels to which it is attached are raised by the hydraulic pump out of the box, and the yarn allowed to drain for a few minutes, after which the travelling crane carries it on to the next box. Into this box the yarn is again lowered by the pump and made to revolve as before, but this time through a solution of the bleaching agent, which immediately re-acting on the carbonate of soda with which the yarn is charged, renders this bleaching agent free from the danger which attends the employment of chlorine, or the ordinary bleaching powder used in the older methods of bleaching. After the yarns have been brought to the desired shade in the solution of HODGES' bleaching agent, they are either removed as before to a new box, and there washed before being soured, or they are thrown into one of the steeps filled with water for the night. These operations are repeated with weaker solutions in the remaining reel boxes, either once or twice according to the shade required.

Mr. HODGES claims as the chief features of his invention,

that it consists, first, in the employment of a bleaching agent which has not hitherto been practically employed, and a cheap method for its production; second, in the preparation of the yarn prior to its being submitted to the action of the bleaching agent, this preparation setting free not only the imprisoned chlorine of the hypochlorite, but also another powerful bleaching agent, oxygen; third, in new and improved machinery, by which the work of bleaching the yarn is greatly shortened; fourth, in doing away with the tedious and expensive operation of exposing the yarn on the grass. If this last were the only feature in Mr. HODGES' invention, the patentee would have greatly improved the process of bleaching; not only, however, does the new process supplant the old long and tedious one, but a great economy of time is additionally gained in other parts of the process; added to these advantages it is stated that a superior finish is given to the yarns, and that in consequence a much greater demand for them has arisen.

Mr. HODGES contends that the absence of caustic lime from his new bleaching compound gives it great advantages over the old bleaching powder, particularly in its application to finely woven fabrics, such as muslins, &c. He also says that fabrics bleached by it receive an increased capacity for imbibing and retaining colouring matter, a fact of considerable importance to the dyer and calico-printer, as they are thus enabled to communicate to the fabrics tints which have heretofore been considered impossible.

The following is the Irish plan of bleaching yarns:—

To Bleach 1,200 *lbs. Weight of Yarn.*

Boil 3 hours in soda-lye.
Reel in solution of hypochlorite of lime, $1\frac{1}{4}$ hours.
Wash for 25 minutes.
Sour in sulphuric acid, $\frac{1}{2}$ hour.

Wash for 25 minutes.
Scald ¾ hour.
Reel in solution of hypochlorite of lime.
Squeeze ¼ hour.
Sour ½ hour.
Wash 25 minutes.
Soap 1 hour.

   This should make the yarns ½ white.

To Bleach 1,200 lbs. Weight of Yarn.

Boil 3 hours in soda-ash of 10 per cent. strength.
Reel 1¼ hours in solution of bleaching powder of 150 per cent. strength.
Wash 25 minutes.
Sour ½ hour in sulphuric acid of 3 per cent. strength.
Wash 25 minutes.
Scald ¾ hour in soda-ash of 3 per cent. strength.
Reel in solution of bleaching powder ½ hour of 40 per cent. strength.
Squeeze ¼ hour.
Sour in sulphuric acid for ½ hour of 1½ per cent. strength.
Wash 25 minutes.
Soap 1 hour, ½ per cent. strength.

   The yarn should now be ½ white.

Scald ½ hour in soda-ash of 1½ per cent. strength.
Spread it on the grass for 24 hours.
Steep in bleaching powder solution 12 hours of 20 per cent. strength.
Wash 30 minutes.
Sour in sulphuric acid ½ hour of 1½ per cent. strength.
Wash 25 minutes.
Soap 1 hour of ½ per cent. strength.

   The yarns should now be ¾ white.

Scald ½ hour in soda-ash of 1½ per cent. strength.

Spread on grass 10 days.
Scald ½ hour in soda-ash of ½ per cent. strength.
Steep 12 hours in bleaching powder solution of 20 per cent. strength.
Wash 30 minutes.
Sour in sulphuric acid ½ hour, of 1⅓ per cent. strength.
 If the yarns are not now full white,
Scald ¼ hour in soda-ash of 1½ per cent. strength.
Steep 12 hours in bleaching powder solution of 10 per cent. strength.
Squeeze ¼ hour.
Sour ½ hour in sulphuric acid of 1½ per cent. strength.
Soap 1 hour, of ½ per cent. strength.
 The yarns should now be full white.

Bleaching Woollen Goods.—In the case of woollen fabrics, the operations of purifying or whitening the wool, beyond the removal of the yolk, are, for the most part, mixed up with the weaving and working of it. The pieces leave the hands of the weaver of a dingy grey colour, loaded with oil, dirt, and dressing. They then pass to the fulling-mill, where they are treated with fuller's earth and soap, often preceded with ammonia or stale urine, after each application of which they are well washed out or scoured with cold water, and are then ready for the dyer. When it is intended to obtain them very white, or to dye them of a very delicate shade, they are commonly sulphured; after which they are washed or milled in cold water for some hours, a little finely ground indigo being added towards the end, to increase their whiteness; an addition also made when the cloth is sufficiently white without the sulphuring process.

 The usual mode of sulphuring woollen goods is to hang them upon pegs or rails, or, in the case of fleece-wool, to spread it about, at the upper part of a close, lofty room or chamber, called a sulphur-stove. In each corner of this room is set a cast-iron pot containing sulphur, which, after the introduction of the goods, is set on fire, when the door at the lower part of the chamber is shut tight and clayed. This is commonly done overnight; and by the morning, the bleaching being finished, the goods are removed, washed, and azured.

Sulphuring, unless very skilfully managed, imparts a harsh feel to woollen goods, which is best removed by a very weak (lukewarm) bath of soap-and-water; but the action of soap in part reproduces the previous yellowish-white tinge. Milling with cold, or lukewarm water, tinged with indigo, is the best substitute.

Raw wool loses from 35 to 45 per cent. of its weight by scouring, and 1 to 2 per cent. more in the subsequent operations of the bleacher; the loss being in direct proportion to the fineness of the staple.

In technical language, the words bleaching, bleacher, bleachery, bleach-works, &c., when employed alone, are understood to have reference only to cotton and linen. This has arisen from the enormous extent of these manufactures, and from the process of bleaching them forming a business entirely distinct from that of weaving, dyeing, or printing them.

Bleaching Silk. — The following is extracted from Mr. Spons' useful volume, " Workshop Receipts." " A lye of white soap is made by boiling in water 30 lbs. of soap for every 100 lbs. of silk intended to be bleached, and in this the silk is steeped till the gum in the silk is dissolved and separated. The silk is then put into bags of coarse cloth, and boiled in a similar lye for an hour. By these processes it loses 25 per cent. of its original weight. The silk is then thoroughly washed and steeped in a hot lye, composed of $1\frac{1}{2}$ lbs. of soap and 90 gallons of water, with a small quantity of litmus and indigo diffused. After this, it is carried to the sulphuring room. Two lbs. of sulphur are sufficient for 100 lbs. of silk. When these processes are not sufficiently successful, it is washed with clear *hard* water, and sulphured again."

Bleaching Feathers.—The process is as follows :—The feathers are first thoroughly washed with soap-and-water, to free them from any oil they may contain. They are next transferred to a bath composed of bichromate of potash dissolved in water, to which has been added a few drops of nitric or sulphuric acid. In this bath they rapidly lose their black, brown, or grey colour, and become almost white. On being removed from this bath they are well rinsed in water, and are then fit to be dyed, even the most delicate colour. Great care is required in the process, as the flue of the feather is apt to be destroyed, if kept too long in the bath. A bleached feather may be readily known by the yellow colour of its stem.

Other methods have been adopted, such as a bath of chloride of lime, peroxide of hydrogen, or sulphurous acid, &c., but the bichromate bath gives the best results.

Bleaching Materials for Paper:—Old rags for the manufacture of paper, and paper-pulp, are almost universally bleached with chlorine or chloride of lime; the former being generally used in France, and the latter in England. The process usually consists in (1) boiling in an alkaline lye to remove grease and dirt, (2) washing, (3) pressing, (4) deviling or tearing up the pressed cake into fine shreds or pulp, (5) chemicking, with agitation, for about an hour, in a clear solution of chloride of lime,* followed by (6) washing, (7) souring

---

* The strength varies with the strength and quality of the rags. From 2 to 4 lbs. per cwt. of rags is a common proportion. For dyed and printed rags as much as 7 or even 8 lbs. per cwt. are

with dilute hydrochloric acid at 1 or 2° Twaddle, or treatment with a solution of some antichlor,* or both, and (8) a final washing and pressing. For the common kinds of paper, the operations included in No. 7 are omitted; but unless the whole of the lime-salt be removed from the pulp, the paper made of it is liable to turn brown and become rotten by age. In some cases rags are bleached before being divided and pulped. Cotton-waste is bleached in a similar way to rags.

In France, the chlorine, in a gaseous form, is passed from the generators into the bleach-cisterns containing the pulp, which in this case must be fitted with close covers.

PRINTED PAPER, as BOOKS, ENGRAVINGS, MAPS, &c. These when stained or discoloured, may be whitened by (1) wetting them with pure clean water, (2) plunging them into a dilute solution of chloride of lime, (3) passing them through water soured with hydrochloric acid, and then (4) through pure water until every trace of acid be removed. This process may be further improved by additionally dipping them into a weak solution of some antichlor, and again washing them, before finally drying them. It is only rare and valuable

---

often employed. It is better, however, to prolong the process with a weaker solution, than to hasten it by using the chloride in excess. Large rectangular cisterns of wood, or of slate, are commonly employed as the bleach-vessels. Cisterns of wood, or brick-work lined with gutta percha or with asphalto-bitumen, are employed in some paper-mills, and answer admirably.

* *Antichlore.*—Among bleachers, any substance, agent, or means by which the pernicious after-effects of chlorine are prevented. Washing with a weak solution of sulphite of soda is commonly adopted for this purpose. Chloride of tin, used in the same way, has been recommended. A cheap sulphite of lime, prepared by agitating milk of lime with the fumes of burning sulphur, and draining and air-drying the product, was patented in England and America by Prof. HORSFORD under the name of "Antichloride of Lime."

original works or specimens of art that are worth this treatment, which, owing to the very nature of paper, requires considerable address to manage. In many cases a sufficient degree of renovation may be effected, by simply exposing the articles, previously slightly moistened, to the fumes of burning sulphur, followed by passing them through a vessel of pure water.

Straw, Straw-plait, and articles made of them, are, on the large scale, usually bleached by (1) a hot steep or boil in a weak solution of caustic soda, or a stronger one of soda-ash, followed (2) by washing, and (3) by exposure to the fumes of burning sulphur. To effect the last, the goods are suspended in a close chamber connected with a small stove, in which brimstone is kept burning. On the small scale, a large chest or box is commonly employed. A piece of brick, or an old box-iron heater, heated to dull redness, is placed at the bottom of an iron crock or earthen pan, a few fragments of roll sulphur thrown on, the lid instantly closed, and the whole left for some hours. Care should be taken to avoid inhaling the fumes, which are very deleterious as well as disagreeable and annoying. Straw goods are also frequently bleached by the use of a weak solution of chloride of lime, or of water strongly soured with oxalic acid or even oil of vitriol, followed by very careful rinsing in clean water; but here, as in the former case, the natural varnish, dirt, grease, &c., must be first removed by alkalies or soap, to enable the chlorine or acid to act on the fibres.

Wax.—Wax is bleached by first melting it at a low temperature in a cauldron, from whence it is allowed to run out by a pipe at the bottom, into a capacious vessel filled with cold water.

This vessel is fitted with a large wooden cylinder, which turns upon its axis, and the melted wax falls upon this cylinder. The surface of the cylinder being always wet, the wax does not adhere to it, but becomes solid, assuming the form of ribbons as it does so, and in this shape becoming distributed through the water in the tub. The wax is then removed and placed upon large frames stretched upon linen cloth, which are supported about 18 inches above the ground, and erected in a situation exposed to the air, dew, and sun. The several ribbons thus placed on the frame should not exceed an inch and a half, and they ought to be so moved about from time to time, as that each part may be equally exposed. If the weather be favourable the wax will become white in a few days. It is again remelted, formed into ribbons, and exposed as before. These operations are continued, until the wax is completely bleached, after which it is melted and run into moulds.

The theory of bleaching, notwithstanding the giant strides of chemistry of late years, remains still unsettled; and hence the processes employed are, for the most part, empirical. It appears probable that chlorine acts by uniting with the hydrogen of the water, or of other compounds present, or probably with that of both, and that it is the oxygen thus liberated, and whilst in the nascent state, that is the true operative agent. Hence bleaching by chlorine, or by the hypochlorites, may be regarded as an oxidation of the colouring matter; but whether the chlorine or the oxygen effects this oxidation is of little practical importance, the result being the same, the destruction of the compound, and the removal of the colour that depends on its existence. It is doubtful whether the bleaching power of sulphurous acid is due to it as an oxidizing or a deoxidizing agent; but the last is probably the case, with a like destruction of the compound constituting the colouring matter. It may be, that sulphurous acid acts as an oxidizer, as when it decomposes sulphuretted hydrogen; or it may act by simply altering the compound by itself combining with it, a view receiving some support from the fact that wool whitened by sulphuring may be restored to nearly its previous colour, by merely treating it with soap or alkalies.

The bleaching power of light depends on its actinic or chemical rays, which, like chlorine, appear to act as an oxidizing agent.

Chlorates, chromates, chromic acid, manganates, &c., have been proposed as bleaching agents for textile filaments and fabrics, but without success or practical advantage. Immersion in water more or less strongly impregnated with sulphurous acid has, however, been successfully substituted for the common sulphuring process, particularly for silk.

To avoid the injury of the goods by sparks, and by drops of water highly saturated with sulphurous acid falling from the roof, Mr. THOM invented a method of passing them rapidly through, or keeping them in constant motion, in the sulphuring chamber. His apparatus is constructed on the principle of the washing-machine, the fumes of burning sulphur being used instead of water.

In addition to the one previously mentioned, the late M. TESSIÉ DU MATHEY proposed a method for bleaching as follows:—He takes about equal parts of permanganate of soda and sulphate of magnesia, and dissolves them in lukewarm water. The tissues, previously freed from grease, are to be plunged into this bath until they are covered with a brown coating. They are then to be placed in a bath of sulphuric acid at 4 per cent., and rinsed after the brown matter is removed. They may be finally passed through sulphurous acid. Mr. RAMSAY's method consists in sprinkling with water equal parts of chloride of lime and sulphate of magnesia when hypochlorite of magnesia is formed. It may be remarked that none of the more modern methods of bleaching, which dispense with the use of chlorine and its compounds, have been found, when reduced to practice, to be cheaper, better, or more advantageous to work than those sanctioned by long experience and use.

CHAPTER II.

DYEING.

Dyeing may be briefly described as the art of tingeing with various colouring matters certain absorbent organic bodies, such as wool, silk, cotton, flax, &c.

Dyeing is an art of great antiquity. Amongst the ancient nations, the Phœnicians, the Romans, and the Egyptians were amongst those who prosecuted it the most successfully. The manufacture of the historic Tyrian purple constituted the principal handicraft of Tyre, and was its chief article of commerce and export. This dye was yielded by a species of mollusk, one single drop only, according to Pliny and Aristotle, being the produce of one animal. Robes dyed with Tyrian purple were worn by one of the Phœnician monarchs as far back as 1500 years B.C. At the commencement of our era, so generally were garments, dyed with the Tyrian purple, spite of their high price, worn by the wealthy Roman citizens, that the Emperor Augustus issued a sumptuary edict, limiting the use of such apparel to himself. From the nature of the dye-stuffs employed by the Romans, it may be inferred that they had attained to some proficiency in the art. They used copperas, native alum mixed with copperas, alkanet root, archil, madder, woad, nut gall, and pomegranate seeds. The ancient Greeks, on the contrary, do not seem to have cultivated dyeing to any extent, since the Athenian people mostly wore dresses of undyed wool. According to Pliny, the ancient Egyptians were expert dyers, and acquainted with the use of mordants, which it would seem they derived from Hindostan. The art was also successfully cultivated by some of the ancient communities of Asia as well as by the ancient Mexicans and Peruvians.

It may be safely said, however, of all these methods of dyeing, as well as of those in vogue prior to the birth of modern Chemistry, that they were carried out by rule of thumb only, and upon no conscious scientific principles.

It was not until about the beginning of the fourteenth century that dyeing seems to have made any progress in Europe. About that period, this branch of industry was prosecuted with remarkable activity in France, and shortly afterwards in Italy. Among the causes that contributed to its spread may be quoted the invention of printing and the discovery of America. This latter event gave a great impetus to the art of dyeing, because of the numerous and valuable dye stuffs the New World sent to the Old. Amongst these tinctorial agents were logwood, Brazil wood, annatto, and cochineal. Owing to these circumstances, and additionally to the publication about the middle of the sixteenth century, of a work on dyeing, by ROSETTI, considerable improvements upon the old methods were introduced, and subsequently adopted throughout England, France, and Germany. The first published English account of the dyeing process appeared in 1664, in SPRATT's "History of the Royal Society."

In the reign of Elizabeth, indigo (which was employed by the ancients as a pigment only), was introduced, or attempted to be introduced, into this country as a dye. It seems hardly credible that the importation of so important and useful a tinctorial substance should have met with an opposition so fierce as it did, and that its use should have been proscribed by Act of Parliament, which Act continued in active operation until Charles II.'s time. Logwood was also excluded by the same legislative enactment. About the middle of the last century, Turkey-red-dyeing was introduced into England and France from India. The first application of iron salts as mordants was made about the middle of the seventeenth century, by DREBBLE, a Dutch dyer. His son-in-law established large dye works at Bow in 1643.

When a fabric is impregnated of a uniform colour over its whole surface, it is said to be simply dyed. If, however, distinct patterns or designs in one or more colours have been impressed upon it, and (as in many cases) portions of it are altogether free from dye, the process by which this is effected, and which is a much more complicated and difficult one than the above, is known as "Calico-Printing," and is described under that head further on.

The process of dyeing would be incomplete unless the fleece,* yarn or cloth after being subjected to it, were more

---

* Wool is sometimes dyed in the flock or fleece before being spun, sometimes as yarn or spun thread, and at others in the finished fabric. It is dyed in the flock or fleece when it is intended for the best varieties of broad cloth. In fleece dyeing there is great liability of the mass becoming so felted together as to interfere considerably with the after operations of carding and spinning. In wool dyeing the use of chipped or splintered dye-woods and lumpy dye-pastes should be avoided, since both these are liable to leave little particles which work into the wool and are very difficult of removal. This inconvenience may be got over by using the dye-stuff in the form of a liquid extract.

Silk is mostly dyed in unspun skeins.

Cotton is rarely dyed in an unspun state, or as cotton wool, but chiefly as yarn or spun thread, arranged either in the "cop" or the *hank*, or skein. In the first case the labour and time, and consequently the expense, of unravelling the skein are saved, but the dye is liable not to permeate the fibre so thoroughly and equally, as when it is in the hank.

or less enabled to retain the tinctorial substance when exposed to certain agencies, such as light, air, washing in soap and water, &c. Hence it is indispensable, both in dyeing simple and calico-printing, there should be as much as possible a permanent adhesion of the colour to the fibre. The varying degrees in which this condition is carried out, constitute the difference between "fast and loose," or "fugitive" dyed goods.

The substances employed by the dyer and calico-printer are obtained from the animal and vegetable kingdom, from the chemical manufacturer, and from the tar colour factory, by far the larger number being procured from the last-named source. These artificial colours, the adoption of which of late years has more or less led to the abandonment of many of the older dye-stuffs, are obtained from what was once a worthless and troublesome waste product in gas making—viz., coal-tar. Hence they are known as "Coal-tar Colours," and they not only, as a general rule, possess much greater brilliancy, depth, and variety of tint than the dyes of natural origin, but some of them have a chemical constitution precisely the same as the colouring principles discovered in certain plants.

The textile fibres used in dyeing, attract and attach to themselves the tinctorial bodies, with very different degrees of force. Wool and silk have a much greater affinity for colouring principles than cotton or linen. Hence dyes, more or less, permanently fix themselves to the former without the intervention of a third substance, but this intermediary (called a *mordant*) is required in dyeing cotton or linen. Colouring principles are divided by BANCROFT into those which do not require a mordant, and are called *substantive* colours, and those which do, and which are termed *adjective* colours. This distinction, however, is a loose one, since the same dye may be a substantive one on one species of fibre, and an adjective one on another. For instance, the aniline dyes are substantive when used with woollen or silken goods, and adjective with cotton ones.

When an infusion of some dye-stuff, such as cochineal or madder, is mixed with alum or acetate of alumina and a little alkali, a precipitate immediately forms, consisting of alumina in combination with colouring matter, constituting a *lake*. It is by a similar reaction, surmised to take place either within or upon the fibres, when these and certain metallic salts and dyes are brought together, that the permanent dyeing of the fibres is effected. The deposition and permanent fixation of the dye to the fibre is sometimes accomplished by other means than by the formation of a lake. We may give the following as examples :—The colouring matters of annatto and safflower being soluble in alkalies, an alkaline solution of these substances is prepared, and the cloth is dipped into it. It has now become saturated with an extremely fugitive colour, but by passing it through acidulated water, the alkaline solvent is neutralized, and the tinctorial matter is precipitated in an insoluble and minutely divided state upon or within the pores, and the fabric becomes permanently dyed. Again, when blue indigo is placed in a vat with certain reagents, it becomes reduced or converted into soluble white indigo, and when woollen or cotton fibre is then placed in the vat, the fibre becomes saturated with the white indigo solution. If now, the cloth be removed from the dye-bath and exposed to the air, the reduced indigo, by the absorption of oxygen, is reconverted into the blue, which is simultaneously fixed to the fibre. Further, when a textile fabric is immersed in an ammoniacal solution of oxide of copper, and afterwards exposed to the air, the ammonia escapes, and leaves the copper oxide deposited in an insoluble condition on the fabric.

Various theories have been propounded as to the nature of the union between the fibre and the colouring matter, or mordant, combined with the colouring matter.

They may all, however, be classed under two divisions:— 1, The Chemical; and 2, The Mechanical Theory.

BERGMAN, CHEVREUL and other chemists, who were the advocates of the first view, maintained that a true chemical combination, confined to the surface, occurs between the fibre and the colouring matter or mordant. But since it has never been shown that the fibre and the colouring principle are united in definite atomic quantities, that neither has lost its distinctive properties by combination, and also that the tinctorial principle can be removed from the cloth to which it has been attached, and dissolved by any liquid which acts upon it, in an uncombined state, whilst the fibre remains intact, the chemical hypothesis has been very generally abandoned.

The chief exponents of the mechanical theory are HELLOT, LE PILEUR D'APLIGNY, PERSOZ, and Mr. WALTER CRUM. HELLOT supposed that the fibre was furnished with pores which expanded under the influence of heat, and thus admitted the tinctorial particles, which were retained owing to the subsequent contraction of the fibre by cold. LE PILEUR D'APLIGNY attributed the different manner in which various fibres act toward the same colouring matter, to the varying size and number of their pores. PERSOZ maintained that the colours or mordanted colours fixed themselves only to the surface of the fibre. Mr. WALTER CRUM, writing on the fixation of colouring matters on cotton, says in effect, either that the pores of the fibre attract the colouring particles from the solution and precipitate them within the fibre, in a manner similar to that of carbon when it absorbs gases, and withdraws solid bodies from their solutions; or that having entered the pores in a state of solution, the colouring particles become fixed by the removal of their solvent, or in some other manner so that they are no longer soluble in water. The opinions of Mr. CRUM are generally accepted at the present day.

It may be mentioned that examination by means of the most powerful microscopes has failed to throw any light upon the nature of textile fabrics that would be of any use to the speculations of the scientific dyer. That an instance of true chemical combination occurs when the mordant, and the colouring principle of the dye are brought together under suitable conditions, there can be little doubt. Evidence of this is afforded by the fact that the colour of the product arising from the reaction is not the same as that of either of its constituents, and that only definite quantities of mordant and colour will give a certain shade; besides which the tints given by any dye-stuff differ with the mordant, and for different strengths of the same mordant. When the dye is brought into contact with a mordant consisting of some metallic salt, the resulting compound may be regarded as an insoluble salt, in which the colour-giving principle performs the part of an acid, and the metallic oxide of a base. Thus Turkey red, and the red colour yielded by a decoction of Brazil wood, when madder and Brazil wood are boiled with a salt of aluminium, may be looked upon, the first as an alizarate, and the second as a brazileate of alumina; whilst the scarlet cochineal colour which is produced when cochineal is boiled with chloride of tin, may be regarded as carminate of tin.

The apparatus employed in mordanting and dyeing textile fabrics, varies with the material and the treatment to which this is subjected. Unspun wool and rags are simply put into the dye-bath, and moved about by a large pole; whilst yarns and woven fabrics are impregnated with the mordant as well as the dye, by means of the "padding" machine. This apparatus consists of a reel (placed above the trough holding the mordant, or dye, as the case may be), around which the cloth, stitched together by the ends, is wound; a roller, which smooths and adjusts the cloth before it enters the trough; a copper cylinder, nearly at the bottom of the trough, under which the fabric is carried from the roller; a half-round polished steel bar to give it equal tension; a pair of padded cylinders, to remove superfluous moisture, and a reel to receive the mordanted or dyed cloth. This apparatus is worked by machinery, and is employed for applying cold solutions of the dye-stuff to cotton goods in the piece. It is also used in many of the operations of bleaching and starching textile fabrics.

In small establishments the goods are dyed by immersion in cisterns, tubs, or vats, and turned by hand; or, if in the form of yarn, they are generally hung on sticks and so suspended in the dye-bath. When the goods are placed in a dye solution, which is required to be kept at the boiling point, or near it, a machine such as that shown in the engraving is used. *a* is a reel turned by steam-power. The cotton-piece, sown together at the ends, is wound round the wooden arms of the reel, as shown in the engraving, taking the direction indicated by the arrows, and after entering the dye-bath, falling on the shelf, *g*, and passing in the course of one revolution under the rollers, *c* and *d*. The orifice, *b*, is for the admission of steam if necessary. The pieces are thus passed through the dye-bath until they have acquired the required depth of shade, during which the cloth is made to turn on the reel with considerable speed and regularity of motion, precautions which ensure a uniform absorption of the dye, and its consequent equal distribution over the surface of the cloth. Care

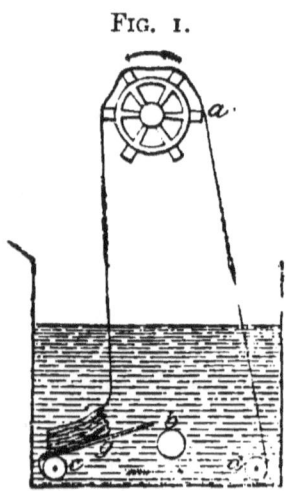

FIG. 1.

must be taken to prevent the goods getting entangled during the "winching," as they are sometimes liable to. They must also coil and uncoil smoothly and easily. Great care is necessary in the selection of the vessels containing the dye-liquid. They should always be made of some material which is unacted upon by the contained fluid. They are variously manufactured of wood, stone, slate, iron, and block tin. For mordants and acid dyes, as well as for bright coloured ones, iron vessels are inadmissible. It is best to have special pans for each dye; and in every case to keep them as nearly absolutely clean as possible. The quality of the water used in dyeng is a matter of first importance, the nearer it approaches distilled water, except when employed for madder dyeing, in purity, the better.* The dyed calico goods of Alsace, and the woollen stuffs of Berlin, in great measure owe their superiority to the excellent quality of the water supply.

Hard waters which contain the salts of lime and magnesia are not well adapted for dyeing. The presence in the water of salts of iron is also injurious. Lime mostly exists in water as carbonate or sulphate, in both which forms it is detrimental to bright colours.

With dark and sombre colours, these salts, unless they are present in the water in large quantities, are not particularly injurious; but the sulphate in small quantity even should be avoided in madder dyeing, or in any case where the dye employed is of a ligneous nature.

Magnesia, where it exists as carbonate, is a very prejudicial ingredient, since it is impossible to dye certain colours with it, and it destroys the appearance of others. It is especially detrimental in madder dyeing. It is rarely present, however, in water in quantity large enough to act

---

* Carbonate of lime, which is a constituent of some varieties of madder root, is added to the water used for the madder bath, when the root and water are deficient in it.

injuriously. Iron, whether in large or small amount, is a most disastrous ingredient. It converts pinks into drabs, and reds into browns and chocolate when present large quantity, it wastes the colouring matter by combining with it, and so removing it from, and preventing its deposition upon the fibre. In ordinary water, the carbonates of potash and soda are seldom in quantity sufficient to act prejudicially. Water containing organic matter is particularly ill adapted for the processes of bleaching and cleansing. Such water is also equally objectionable for dyeing. The method usually adopted to get rid of it is filtration and exposure to the air and light. Mr. CROOKES says another method consists in adding successively to each 1,000 cubic centimetres (35,316 cubic feet, or 220,096 gallons) of water, which should be contained in a suitable tank, about 6 lbs. of dry perchloride of iron, and 186 lbs. of crystallized carbonate of soda, both previously dissolved in as pure water as can be obtained, and to the volume of about 220 gallons (35 cubic feet). When the iron and soda are added to the water, the whole should be vigorously stirred.

It has been already stated that the contents of the dye-vessel are sometimes heated by means of steam blown into them. In some dye-works the pans are exposed to the naked fire; at others they are raised to the required temperature by steam coils; or in some establishments, they are jacketted, and so heated.

Before the goods (provided they have been mordanted) are passed through the dye-beck, it is necessary first, to fix or attach the mordant* to them; and second, to remove any superfluous adhering mordant.

---

* The mordant is sometimes applied before the dye, sometimes after, and sometimes simultaneously. This latter method, however, does not admit of extensive application, since in most cases the

A general description of these two processes will be found further on.

We append formulæ for plain dyeing:—

COTTON.

Black.—1. The goods, previously dyed blue, are steeped for about 24 hours in a decoction of gall-nuts or sumach, then drained, rinsed in water, and passed through a bath of acetate of iron for a quarter of an hour; they are next again rinsed in water, and exposed for some time to the air; after which they are passed a second time through the bath, to which a little more iron-liquor is previously added. The whole process is repeated, if necessary, according to the intensity of the shade of black desired.

2. The goods are steeped in a mordant of acetate of iron, worked well, and then passed through a bath of madder and logwood for 2 hours. Less permanent than No. 1.

About 2 oz. of coarsely powdered galls, or 4 oz. of sumach, are required for every pound of cotton, in the process of galling. The first should be boiled in the water, in the proportion of about $\frac{1}{2}$ gal. of water to every 1 lb. of cotton. The sumach-bath is better made by mere infusion of that dye-stuff in very hot water.

3. For 10 lbs. of Cloth.—The goods are put into a boiling bath made of 3 lbs. of sumach, and allowed to steep, with occasional "working," until the liquor is perfectly cold; they are next passed through lime water, and, after having

---

products are insoluble lakes, which give rise to loose and inferior colours. Again, the fixing of the mordant in calico-printing involves a number of processes which are not required in piece-dyeing. So, again, whilst wool is almost always dyed at a boiling temperature, in most cases the dyeing of cotton is performed either in the cold or at a temperature of from 90° to 100° Fahr. (32·2° to 38° C.).

drained for a few minutes, immediately transferred to and worked for an hour in a warm solution of 2 lbs. of copperas; after free exposure to the air for about an hour they are again passed through lime water, and, after draining, "worked" for an hour in a bath made with 3 lbs. of logwood, and 1 lb. of fustic; they are then "lifted," and ¼ lb. of copperas being added, they are returned to the bath, "worked" well for about 30 minutes, and finished. Good and deep.

Instead of copperas, iron-liquor may be used, observing to take 1½ pint of the latter (of the ordinary strength) for every 1 lb. of the former ordered above.

4. *For 40 lbs. of Cotton.*—Boil or scald, sumach 10 lbs. Let the cloth or yarn remain in this for 18 hours; wring out; run through acetate of iron at 40° Twaddle four turns or for ½ an hour; wring out; repeat and thoroughly wash in three waters. Next boil together 8 lbs. logwood and 1 lb. fustic; put off the boil and enter, or the clear portion of the decoction may be decanted into another vessel; one run, continue ½ an hour; wring out; repeat; sadden with 1 lb. of copperas; 2 runs; wash and dry.

5. *For 100 lbs. of Cotton.*—Steep in a decoction of 30 lbs. of sumach at a boiling heat, and let stand till quite cold, then pass through lime water, and then directly after, work for 1 hour in a solution of 20 lbs. of copperas. After this expose to the air for 1 hour; then pass a second time through lime water, and wash and work for 1 hour in a bath of 30 lbs. logwood and 10 lbs. fustic; lift and add 2 lbs. copperas; and work ½ hour longer, and finish.

6. *Common Black.* 5 pieces = 75 lbs. are padded through acetate of iron (iron liquor) at 8° Twaddle, dried and afterwards passed through lime water (milk of lime); afterwards washed, then dyed with 35 lbs. of ground logwood and 3 lbs. of fustic extract at 48° Twaddle; in this they are worked

for ½ an hour at boil; then winched, rinsed, and dried. They are further run through a little starch water containing a small quantity of soap, and finally dried for finishing.

7. *Good Common Black* (Carlisle Finish).—7 pieces = 85 lbs. are worked in the jigger, cold for 6 ends, and afterwards passed through a water-mangle to squeeze out a large portion of the liquor; then dried; they are then padded in acetate of iron at 8° Twaddle, and dried out of it; afterwards again entered into the jigger, which is charged with sufficient water and 5 lbs. of chalk (carbonate of lime); give two ends; then wash, and afterwards dye with 48 lbs. ground logwood and 3½ lbs. fustic extract at 48° Twaddle; work in the jigger for 45 minutes at boil; wash and dry.

8. *Chrome Black* (Italian Black).—6 pieces satin (cotton) = 108 lbs. Work in jigger containing 20 lbs. of sumach (Palermo), and 20 lbs. of myrabolams, in as little water as possible, and at boil for 7 to 8 ends; then run off the liquor and recharge the jigger with 15 gals. water and 5 lbs. sulphate of copper, cold; give 4 ends in this; again wash well, and recharge the jigger with bichromate of potash, at, say, 2° Twaddle; give 2 ends cold, and then 3 ends at boil; again wash and afterwards dye in the jigger, it being recharged with 72 lbs. ground logwood and 4½ lbs. fustic extract at 48° Twaddle; work backward and forward at boil for 1 hour; then rinse in a weak solution of soda or potash, say, 8 oz. to 20 gals. water; wash and dry.

9. *Aniline Black with Vanadium.* (PINKNEY's Patent.)—Take of hydrochlorate of aniline, 150 parts; salt of vanadium, ⅛ part; chloride of nickel, 20 parts; potassium chlorate, 100 parts; water, 2,500 parts. The yarns, after being steeped in this mixture, may be dried either hot or cold.*

---

* In practice the salt of vanadium may be considerably reduced in amount, and the chloride of nickel omitted.

10. (DE VINANT): For Cotton Yarns.—This process must be performed with great care, otherwise the colours will be uneven and clouded. The cotton yarn, first well boiled out, receives 7 turns in a bath, consisting of 200 grammes of sulphate of copper for every kilo of material dissolved in water, which has been slightly acidulated with hydrochloric acid. It must be then thoroughly wrung out. It is next subjected to 5 turns, in a bath containing 50 grammes of hydrosulphate of soda to a litre of water, the temperature of which is at 50° C. (122° Fahr.); after which it is rinsed. It then receives 7 turns in a bath of 10 litres of water, 180 grammes of chlorate of potassium, and 170 grammes of chloride of ammonium, dissolved by means of heat, to which is added 480 grammes of chloride of aniline. It is then stretched out very regularly in a drying-room, kept at 48° C. (118·4° Fahr.), for 48 hours. After this it is turned 4 times at 30° C. (86° Fahr.), in a bath which contains 1 gramme of bichromate of potassium to the litre; after which it is well rinsed and dried. Should the blacks have acquired a reddish tone, they should be passed through a bath consisting of 1 litre of bleaching liquid at 6° B., and 100 litres of cold water.

11. For 100 lbs.—Mix 6 lbs. 9 oz. of aniline oil with 8 lbs. 12 oz. of hydrochloric acid at 32° Twaddle. When the mixture has cooled, add to it a solution consisting of 4 lbs. 6 oz. of chlorate of potassium and 66 parts of water; and then 42¾ pints of solution of chloride of iron at 32° Twaddle. Steep the bleached yarn from 8 to 10 hours in this liquid, previously diluted with water at about 100° Fahr. (38° C.); remove, and then place it in a solution of soda at 23° Twaddle. Then wash or steep for ½ hour in a dye-beck, made up with 66 parts of water and 7 oz. of chromate of potassium, at about 112° Fahr. (44·4° C.). This prevents the dye turning green afterwards. Wash and pass the yarn through a mixture

of 17¼ oz. of emulsive oil (such as is used in Turkey red dying), 2 lbs. 3 oz. potash, and 66 parts of water, and dry immediately.

N.B.—Mr. JAMES CHADWICK writes us :—" To produce plain aniline black, we print a colour composed of chloride of aniline, chlorate of soda and vanadiate of ammonia. This colour we print on with what is called a " pin pad roller," and on both sides of the fabric, thus completely covering every thread. When dried as usual after printing, we age through an aniline ager at from 180° Fahr. (82·2° C.), and 170° moisture; afterwards pass through a solution of bichromate of potash at 2° Twaddle, cold. Wash well and slightly soap; wash and dry. I believe most of the plain aniline blacks are done this way." Mr. CHADWICK adds :— " I have not seen a good dyed aniline black. The fibre of the fabric is always more or less injured, and the colour always turns green in the presence of sulphuric acid, which abounds, as you know, in towns where much coal is consumed."

Brown.—12. *Bismark Brown.*—For 10 lbs. of cloth or yarn, work in a hot decoction of ½ lb. of sumach for ½ hour; wring out and work for 20 minutes in a solution consisting of 4½ oz. of stannate of sodium, and then thoroughly wash from this. Dissolve 4 oz. of Bismark Brown in the dye beck or boiler, and work the goods in this for ¼ hour, at 120° Fahr. (48·8° C.), or at a heat about as hot as the hand can bear; then wring out to dry. If a redder shade than this preparation will yield, be desired, a little red liquor must be added to the dye; if a yellower tint be required, this may be got by the addition of a little fustic.

13. *Catechu Brown.*—For 10 lbs. of cloth or yarn. The goods must be worked at a boiling heat for
steeped for several hours, if the liquor is allowed to cool,

in 2 lbs. of catechu,* and then worked for ½ hour in a hot solution of potassium bichromate, and washed from this in hot water. A little soap added to the washing water improves the colour.

14. *Dark Brown.*

    7 pieces = 84 lbs. Work in jigger, charged with
   12 gallons boiling water, and
   20 lbs. catechu,
    5 lbs. sumach (Palermo), and
    3 lbs. sulphate of copper; give in this 5 ends, then recharge with 1 gallon acetate of iron, at 12° Twaddle, and 5 lbs. sulphate of iron cold, and 12 gallons water cold. Give 4 ends, and afterwards wash again, and recharge jigger with 12 galls. water, boiling, and 3 lbs. bichromate of potash; give 4 ends, then wash and dry.

15. *Medium and Light Brown* can be obtained by decreasing the quantities of ingredients.

16. *Bright Brown.*—Cotton.

   72 lbs. cloth in jigger, with
    9 gallons boiling water in which dissolve;
    6 lbs. catechu; give 4 ends; recharge jigger with
    9 gallons hot water, 140° Fahr. (60° C.), and
    8 oz. bichromate of potash, and
    2 oz. sulphate of copper. Give 3 ends in this, and wash again; recharge jigger with
   10 gallons warm water, *blood* heat;
    4 oz. Bismark brown powder, and
    8 oz. protochloride of tin, at 120° Twaddle, give 4 ends in this, and afterwards wash and dry as usual.

This gives a beautiful colour.

---

* Boil 1 lb. of catechu in 7 or 8 lbs. of water until dissolved, then add 2 oz. nitrate or sulphate of copper, and stir; after which it will be ready for use.

17. Prussian Blue.

7 pieces = 84 lbs. Work in jigger, containing
15 gallons cold water,
5 quarts nitrate of iron, at 84° Twaddle,
1 pint protochloride of tin at 120° Twaddle; give 4 ends, afterwards wash in cold water, and recharge jigger with
15 gallons water, in which is dissolved
5 lbs. yellow prussiate of potash, and
1 gill sulphuric acid, at 170° Twaddle.

Give four ends, wash and dry.

18. Aniline Blue.

7 pieces = 84 lbs. Work in stannate of soda at 4° Twaddle, 4 ends, then in sulphuric acid, 1° Twaddle, 4 ends, and afterwards 4 ends in water, then recharge jigger with 8 oz. cotton aniline blue, and 8 oz. alum in twelve gallons water, five to six ends, wash and dry.

19. Navy Blue.

7 pieces = 84 lbs. Work in jigger charged with
10 lbs. sumach,
10 lbs. ground logwood,
15 gallons boiling water, give 4 ends, then recharge jigger with 4 quarts nitrate of iron at 84° Twaddle, and 9 gallons of cold water, in which give 4 ends, and afterwards wash, then recharge with 15 gallons water, 4 lbs. yellow prussiate of potash, and half gill sulphuric acid at 170° Twaddle, give four ends, and wash in cold water, recharge with 15 gallons water "cold," and 6 ozs. BB violet crystals (coal-tar), give five ends in this and dry.

20. Bright Green.

7 pieces = 84 lbs. Work in jigger charged with 14 lbs. sumach boiling, give 4 ends, recharge with 15 gallons

water, 3 gills protochloride of tin at 120° Twaddle, and 4 oz. of tartrate of antimony (tartar emetic) in cold water, give 4 ends and afterwards wash. Again recharge with 10 gallons of cold water, and 14 oz. of malachite green (coal-tar), and from 4 to 7 lbs. of fustic extract at 48° Twaddle, according to shade of green required. Give 4 ends in this, wash and dry.

21. **Slate** (*Silesian*).

    7 pieces = 84 lbs. worked in jigger charged with
  15 gallons logwood liquor (at 1 lb. per gallon).
  15 ,, water at 120° Fahr. (49° C.), give 4 ends, then recharge jigger with 30 gallons cold water, and dissolve in it 4 lbs. sulphate of iron, give 3 ends, then dry.

If the shade of slate is too red the addition of a little fustic extract corrects that, of course.

22. **Dark Mauve or Violet.**

7 pieces = 84 lbs. worked in jigger charged with 12 lbs. of ground sumach and 12 gallons of hot water, give 4 ends, and afterwards give 6 ends in jigger charged with 3 gills of protochloride of tin at 120° Twaddle, and 12 gallons of cold water, afterwards wash, and then charge jigger with 9 gallons water (hot), and 8 oz. RR violet crystals, give 4 to 6 ends, and wash and dry as usual.

N.B.—The above gives a deep shade of mauve. If a medium or lighter hue be required, reduce the quantities of ingredients used.

If a blue mauve be required, use 6 B violet crystals instead of the RR, and so on for any other tone or shade of mauve.

23. **Chamois.**

    5 pieces = 80 lbs. cloth.
  12 gallons water, at 100° Fahr. (37·7° C.); jigger charged with

DYEING.

    3 pints catechu 4° Twaddle, give 1 end, then add

    2 pints catechu at 4° Twaddle, give 4 ends more, recharge jigger with same water, and

    3 quarts bichromate potash (1 lb. per gallon), give 1 end, then add

    2 quarts more bichromate potash (1 lb. per gallon), give 4 ends—wash.

24. Canary.

    5 pieces = 80 lbs. cloth, jigger charged with

    12 gallons cold water,

    2 pints fustic extract at 48° Twaddle, give 1 end, then add

    1½ pints fustic extract at 48° Twaddle, give 4 ends, and recharge jigger with the same quantity of water, and 3 lbs. of alum, give 4 ends, and afterwards 2 in water, and then the goods are ready.

25. Claret.

    72 lbs. cloth, jigger charged with

    12 gallons of hot water at 120° Fahr. (49° C.), and

    10 lbs. sumach, and

    10 lbs. ground logwood, give 5 ends in this, then add to it

    5 gills protochloride of tin at 120° Twaddle, give 4 ends more and wash, recharge jigger with

    12 gallons hot water, 120° Fahr.,

    10 lbs. ground logwood, and

    5 lbs. peachwood, give 4 ends in this and afterwards add to same charge,

    8 oz. ground alum, dissolve, and give 2 more ends, wash as usual.

26. Drab (*Silesian*).

    7 pieces = 84 lbs. jigger charged with

    10 lbs. sumach.

1 lb. ground logwood,
1 lb. ground fustic,
4 oz. annatto,
20 gallons water at 170° Fahr. (76·6° C.), give 4 ends. Then recharge with 12 gallons water in which is dissolved 3 lbs. sulphate of iron.
1 pint nitrate of iron at 84° Twaddle.
$\frac{1}{3}$ gill sulphuric acid at 170° Twaddle, give 2 ends, and wash and dry.

27. Bright Drab.

72 lbs. cloth, jigger charged with
$1\frac{1}{2}$ lbs. sumach.
9 gallons hot water at 120° Fahr. (49° C.), give 4 rounds, then add
4 oz. sulphate of iron, give 4 more rounds. Again charge with 9 gallons of hot water at 120° F.
$2\frac{1}{2}$ lbs. fustic, and
5 oz. extract of indigo, give 4 rounds and wash.

When cotton is dyed with the coal-tar colours, various mordants differing with the specific colours are employed. We give a list of the most important of these mordants:— Sumach, tannic acid, alumina (acetate), glycerine, oleine, stannate of sodium. Stannic and tannic acids, together and in conjunction with alumina, are frequently used, and are the most potent and effective. Acetate or nitro-acetate of chromium are also good mordants for the darker colours. Lead (acetate), Glauber salts, and arsenic are likewise employed, but not quite so extensively, at least lead and arsenic are not. Further on will be found representative formulæ for the coal-tar colours, by following which, and having recourse to the above-mentioned mordants (which are not mentioned in the formulæ referred to) almost every aniline colour can be made available.

28. Bright Rose Pink.

7 pieces = 84 lbs. worked in jigger, with 12 gallons water, cold, containing 3 lbs. acetate of lead, give 3 ends, and afterwards 2 ends in lime-water (milk of lime), and then 4 ends in cold water to remove the lime. Then recharge the jigger with 10 gallons water, and 6 to 8 oz. of eosine, give 5 ends at about 90 to 100° Fahr. ($32·2°$–$37·7°$ C.). Afterwards pass through mangle, and dry.

N.B.—Erythnosine may be used instead of eosine, when a yellower shade of rose is required.

These two red-colouring matters can also be fixed by using common salt (chloride of sodium), instead of lead and lime-water, but the colours are not so full in tone.

These pinks are not permanent.

29. Light Pink (*Magenta*).

7 pieces = 84 lbs. worked in jigger, charged with 12 gallons of cold water, and $1\frac{1}{2}$ lbs. stannate of soda, give 4 ends, and then 4 ends in jigger with 1 gill of sulphuric acid, at 170° Twaddle, and 20 gallons of water. Then wash and recharge jigger with 3 pints of magenta liquor (at 1 oz. of crystals per gallon), and 9 gallons of cold water, give 4 to 6 ends, mangle and dry as usual.

30. Safflower Pink.

7 pieces = 84 lbs., worked in jigger with 20 gallons water at blood heat, and 9 oz. safflower liquor of commerce, give 3 to 4 ends, then add to the same bath $\frac{1}{2}$ gill sulphuric acid at 170° Twaddle, give 2 ends more in this liquor, which is to precipitate the colouring matter into the fibre of the cloth. Wash, mangle, and dry.

31. Barwood Red.

72 lbs. cloth padded through stannate of soda, at 12° Twaddle. Then passed through sulphuric acid at 2° Twaddle, and washed, then in jigger charged with 9 lbs. sumach, and

60 lbs. barwood; run in this charge at boil for 1½ hours. Afterwards wash.

32. Rich Orange.
72 lbs. cloth; jigger charged with
12 gallons hot water, 120° Fahr. (49° C.), and
12 oz. soda ash,
4 lbs. annatto, dissolve and add
4 lbs. turmeric—give 4 ends in this, then add to same
8 oz. (fluid) of sulphuric acid at 170° Twaddle, give 2 ends in it, and afterwards wash. This is much cheaper than chrome orange, and good.

33. Bright Yellow (*Turmeric*).
72 lbs. cloth, say 6 pieces, 70 yards. Run the goods in jigger in hot water to thoroughly and evenly wet them, then to 20 gallons of hot water at 140° Fahr. (60° C.), add 7 lbs. turmeric, give 4 ends, then add to the same liquor 4 fluid ounces of sulphuric acid at 170° Twaddle, and give the goods 2 ends more. Afterwards wash, mangle, and dry.

34. Chrome Yellow.
72 lbs. cloth. Pad through acetate of lead, at strength of 8 oz. per gallon, then pass into jigger charged with limewater, wash in water, and recharge jigger with 9 gallons of water, in which is dissolved 1½ lbs. bichromate of potash, give 4 ends, and afterwards wash. Should the yellow be rather too much of a gold colour, 1 or 2 ends in weak hydrochloric acid will bring it back.

DYEING WOOL AND SILK.

Since the discovery of the coal-tar colours, which are now so extensively used for dyeing wool and silk, most of the older colouring matters are sparingly used, except for the production of dark colours, such as chocolates, dark browns, maroons, and blacks, as well as some intense colours. The coal-tar colours are much easier of application, many of them are cheaper than the old dye stuffs, and most of them require but little mordant, and for silk and wool, some of them, indeed, none.

Their employment, therefore, entails less labour, and they generally give brighter tints and more regular results. The yarn or cloth must, of course, be scoured or bleached before dyeing.

The woollen goods are not worked in jiggers, but in dye vessels, the quantities of water in each case being as much as easily to work the cloth in.*

SILK.

1. **Peacock Blue.**

    80 lbs. silk,
    1 pint sulphuric acid at 170° Twaddle,
    10 oz. methylin blue crystal dye at 120° to 160° Fahr. (49°–71° C.), usual manipulation.

2. **Peacock Blue.**

    80 lbs. cloth or yarn,
    3 oz. biborate of soda (borax),
    11 oz. peacock blue crystals, enter at 140° Fahr. (60 C.), and bring to boil in twenty minutes.

WOOL.

3. **Black.**

    5 pieces = 100 lbs. cloth,
    150 gallons hot water (usual vessels),
    $3\frac{3}{4}$ lbs. bichromate of potash,

---

* In wool dyeing, the goods should in every case be entered just below boiling point, and be well saturated with the dye-liquor, before the bath is raised to boiling.

1 pint sulphuric acid, at 170° Twaddle, enter and work at boil for 1 hour, then wash in cold water and dye in

150 gallons hot water, and

37 lbs. ground logwood,

7 lbs. fustic, chips or ground, enter at boil, and work for 1 hour at boil, when add to the bath

1¾ lbs. sulphate of copper. Continue to work in this bath at boil for 30 minutes. Wash in cold water and dry.

Some dyers use a little alum to the mordant bath.

4. Superior Jet Black.

5 pieces = 100 lbs. cloth; dyeing vessel containing about

150 gallons hot water, in which dissolve

3 lbs. bichromate of potash; work in this at boil for 1 hour, afterwards wash in cold water. Then in same vessel again, charged with

150 gallons hot water, add

34 lbs. ground logwood, and

7 lbs. fustic, ground or chipped; enter just below boiling, and immediately raise to boil, and work in it for 1 hour and 20 minutes. Wash in cold water and dry.

5. Superior Blue Black.

5 pieces = 100 lbs. cloth, in an ordinary dyeing vessel containing about

150 gallons water, dissolve

3 lbs. bichromate of potash, work in this at boil for 1 hour, afterwards wash in cold water. Then in same vessel again, charged with

150 gallons hot water, add

30 lbs. ground logwood; enter same goods, at 180°

DYEING.

Fahr. (82° C.), and then raise the bath to boiling, in which work the goods for 1¼ hours. Then wash in cold water and dry.

6. Dark Brown.

    5 pieces = 100 lbs. cloth,
    20 lbs. turmeric,
    4 lbs. extract of indigo,
    15 lbs. cudbear,
    2 pints sulphuric acid, 170° Twaddle,
    10 lbs. sulphate of soda, enter at boil, and work for about 90 minutes.

7. Claret.

    5 pieces = 100 lbs. cloth,
    80 to 100 gallons water,
    30 lbs. cudbear,
    3 gills sulphuric acid, at 170° Twaddle,
    1 lb. extract of indigo,
    2 oz. magenta (acid) crystals, heat up to near boil before entering, and work for 75 minutes at boil.

8. Dark Chocolate.

    80 lbs. cloth or yarn,
    80 to 100 gallons water,
    3 lbs. bichromate of potash,
    15 lbs. peachwood, ground,
    3¼ lbs. logwood, ground,
    1½ lbs. tartrate of potash, usual manipulation, boil 30 to 40 minutes.

9. Red Drab.

    5 pieces = 100 lbs. cloth, usual water in vessel,
    13 lbs. sulphate of soda,
    4 pints sulphuric acid, 170° Twaddle,

6 lbs. alum,
12 oz. bitartrate of potash,
6 oz. fustic extract at 48° Twaddle,
6 oz. cudbear,
1 oz. extract of indigo, enter at boil and work 45 minutes.

10. **Green.**
3 pieces = 100 lbs. cloth,
80 to 100 gallons water,
10 lbs. alum.
1 pint sulphuric acid, at 170° Twaddle,
10 lbs. extract of indigo.
1¼ lbs. picric acid.
Boil and enter, and work for 90 minutes.

11. **Dark Green.**
5 pieces — 100 lbs. cloth,
2 pints sulphuric acid, 170° Twaddle,
5 lbs. sulphate of soda,
15 lbs. extract of indigo,
1½ lbs. picric acid,
5 lbs. cudbear.
Enter at boil, and work for 90 minutes.

12. **Olive.**
5 pieces = 100 lbs. cloth. Usual quantity of water in vessel,
16 lbs. sulphate of soda,
2½ lbs. alum,
8 oz. bitartrate of potash,
2½ pints sulphuric acid, 170° Twaddle,
8 oz. picric acid,
8 oz. extract of indigo,
1½ oz. cudbear.
Enter at boil, and work 45 to 60 minutes.

13. Salmon.

5 pieces = 100 lbs. of cloth; dye vessel; water usual quantity,
2 pints protochloride of tin, 120° Twaddle,
4 oz. flavine,
6 oz. cochineal,
4 lbs. bitartrate of potash.

Enter at boil, and work for 40 minutes. Wash.

14. Scarlet (*Cochineal*).

5 pieces = 100 lbs. of cloth, say wool,
100 gallons boiling water,
10 lbs. bitartrate of potash, and
6 pints protochloride of tin at 120° Twaddle. Work for 30 minutes, then wind on winch, and add to the same bath
15 lbs. cochineal,
14 ozs. fustic. Enter the goods again, and work for 15 minutes at 180° Fahr. (82° C.), then raise the bath to boiling, and work at boil for 1 hour. Wash and afterwards dry.

15. Scarlet.

75 lbs. cloth or yarn,
1¾ lbs. eosine dissolved in the bath, say at 120° Fahr. (49° C.), add
3 gills sulphuric acid at 170° Twaddle. Enter the goods, say at 140° to 145° Fahr. (60°–63° C.), and gradually bring to boil in from 15 to 20 minutes, and take out.

16. Crimson.

5 pieces = 100 lbs. cloth; dye vessel; water usual quantity,
4 pints protochloride of tin 120° Twaddle,

10 lbs. bitartrate of potash,
30 lbs. cochineal (ammoniacal). No. 17 (*below*).
Work first for 30 minutes in the tin and bitartrate of potash, before adding the ammoniacal cochineal; then after that is added, work for an hour at boil; wash.

17. **Ammoniacal Cochineal.**
18 lbs. cochineal,
14 lbs. liquid ammonia,
2 gallons of water,
Steep for 12 hours in a closed vessel; when it will be ready for use.

18. **Yellow.**
80 lbs. cloth or yarn,
8 lbs. Glauber salts,
1 gill sulphuric acid, 170° Twaddle,
5 oz. aniline yellow.
Usual manipulation.

19. **Yellow.** (*Bark*).
80 lbs. cloth or yarn,
5½ lbs. ground bark,
4¼ lbs. bitartrate of potash,
4¼ quarts protochloride of tin, at 120° Twaddle.
Enter at 140° Fahr. (60° C.), and boil 35 minutes.

20. **Yellow.**
5 pieces = 100 lbs. cloth; dye vessel; water usual quantity,
2 pints protochloride of tin, at 120° Twaddle,
12 lbs. flavine,
5 lbs. oxalic acid.
Enter at 200° Fahr. (93·3° C.), and work quickly for 2 hours; wash.

CHAPTER III.

CALICO-PRINTING.

This branch of dyeing consists in topically printing upon textile fabrics, such as cotton, linen, woollen, silk and mixed goods, figures and designs mostly in two or more colours, upon a white or coloured ground.

It is believed that the ancient Egyptians, who, according to Pliny, practised the art of calico-printing with considerable success, acquired their knowledge from India, which country formerly enjoyed a great renown for the manufacture of its cotton cloth, and its printed calicoes.*

The Dutch are supposed to have been the first who attempted to introduce the art of calico-printing, as practised in India, into Europe; at what period is uncertain. It was possibly from Holland that the art reached Germany, for we find that, towards the end of the seventeenth century, Augsburg was famous for the excellence of its linen and cotton fabrics. Dr. Ure states that calico-printing was probably introduced into England by some Flemish emigrants about the year 1676. Shortly after its introduction, several printing works were set up in this country, with the object of supplying the London shops with chintzes, the importation of which had been prohibited by Act of Parliament in 1600. This introduction of calico goods, whether of native or Indian

---

* The word calico is derived from Calicut, a town in India where these industries were carried on.

origin, met with the most unqualified opposition from the silk and woollen weavers, who in 1680 attacked the India House, for having been instrumental in shipping some Malabar chintzes. Intimidated by the Spitalfield weavers, the Government in that year not only excluded from the English markets the beautiful printed cotton fabrics of Calicut, but passed a law by which the weaving of all printed calicoes whatsoever, whether of foreign or domestic origin, was interdicted, and it was only till after a lapse of 140 years, that printed calico goods were allowed to enter into equitable competition with other fabrics.

In 1738, calico-printing was introduced into Scotland, and in 1764 into Lancashire, the first Lancashire calico-printer being Mr. CLAYTON, who erected works at Bembridge, near Preston. The process was mostly performed by means of engraved blocks applied to the cloth by hand, till the year 1785, when the cylinder printing machine, now almost universally adopted, was invented by a Scotchman of the name of BELL.

In 1746, the first calico works in Mulhausen were established by KOECHLIN, whose descendants are the representatives of the old house which has rendered the printed calicoes of that city so renowned.

The Chinese practised the art many centuries before it was known in Europe.

The process of calico-printing is performed either by means of wooden stamping blocks, on which the patterns are cut in relief ; but more frequently by the aid of engraved copper cylinders. In the former case the work is performed by hand ; in the latter by machinery. The blocks* are of pear-

---

* When a pattern is to be several times reproduced on the block, as many castings of it as are required, are taken in type metal, and afterwards arranged and nailed to a plain block of wood.

tree, box, fir or sycamore wood, with sometimes a piece of copper wire, forming the design, let into them. By means of sharp pins let into the corners of the block, the printer can place it in its exact position on the fabric. The printer dipping the raised part of the block into the mordant, which stands near him (for every different colour he requires a different mordant), stamps it sharply on to the fabric, taking care in so doing to press it evenly and equally on to the cloth, stretched on a flat table covered with a blanket.*

A method of block-printing by machinery was invented by PERROT, of Rouen, in 1833. The "Perrotine," as the apparatus by which this process is carried out, is called, consists of three blocks either of wood or fusible metal, on the faces of which are the patterns or designs in relief. These blocks are fixed at right angles to each other, in a powerful iron frame, and in close proximity to a revolving iron prism covered with cloth. When the fabric, which is drawn through the apparatus by a winding cylinder, passes between the iron prism and the engraved blocks, these latter are pressed against it by means of springs, and print the design upon it in succession. After each impression fresh mordant is applied to every block from a woollen pad, which is smeared over with the paste by means of a mechanical brush. The Perrotine is employed in the French and Belgian factories, but not in the English, and effects a great saving of time and labour over hand-block printing; what required twenty men and twenty children by this latter method, being accomplished by the Perrotine by one man and two children.

In certain cases, such as in printing woollen and mousseline-

---

* Block-printing is also employed for introducing colours into a fabric after it has been subjected to cylinder-printing and dyed.

de-laine goods, block work has been very generally superseded by cylinder printing. By the latter method the mordant is applied to the cloth by means of copper cylinders, whereon the design or pattern is engraved. These cylinders are set in a strong iron framework, and form part of a large apparatus essentially consisting, besides the cylinders, of an iron drum and a colour cylinder. The iron drum is encased in woollen cloth or felt. The annexed drawing,

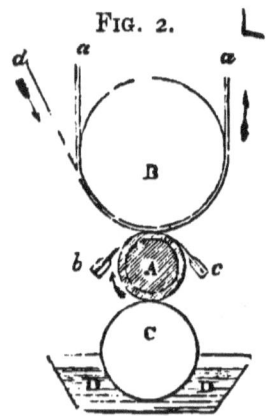

FIG. 2.

which represents a section of the machine, will explain the manner in which it works. B is an iron drum turning on supports fixed in the end framework of the machine. A the engraved cylinder, and C the colour or mordant cylinder covered with woollen cloth. C dips into a trough, D D, filled with the mordant. d is the fabric to be printed. a a is an endless web or blanket passing around B. The engraved cylinder A is mounted on an axis or iron mandrel turned by steam or water power. Being pressed against the surface of the drum B, by means of weights or screws, in revolving, A imparts a rotary movement to B, which carrying the fabric d with it in the direction of the arrows, brings it into contact with the engraved cylinder, and in so doing causes the pattern or design to be printed on it. As it revolves, A imparts a similar movement to the mordant cylinder C, and at the same time robs this of the mordant which it has taken up from the trough D D. Any excess of mordant or colour beyond that required to fill up the engraved lines is scraped off from the face of the cylinder A before this reaches the fabric by b, a sharp metal instrument, called the "colour doctor;" opposite which is a similar contrivance c, for removing any loose threads from the roller, called the "lint doctor." When more than one colour has to be printed, additional cylinders, mordant rollers, colour boxes, &c., are employed. Some apparatus are made to print fabrics in as many as twenty-four different colours.

FIG. 3.

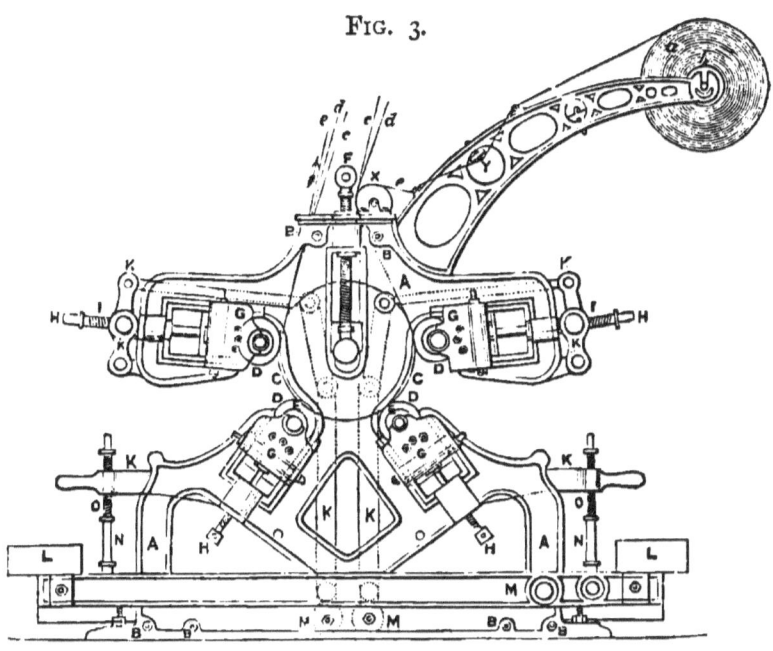

The above engraving represents an end elevation of a four colour calico-printing machine. A is the framework of cast-iron fixed to a corresponding framework by strong bolts B. C is the drum, the diameter of which varies with the fabric to be printed; D the engraved cylinders. E are mandrels made of wrought-iron on which the engraved cylinders are forced by a screw-press. The cylinders D are made with projecting pieces inside, extending all the width of the roller, an arrangement which causes it to revolve with the mandrel, without slipping. The drum, C, rests with its gudgeons on bearings which can be shifted up and down in slots of the side cheeks A. These bearings are suspended

F

from massive screws F, which turn in brass nuts fastened to the framework A. By means of these screws the upward pressure of the two lowest engraved cylinders is counteracted. G G are sliding pieces, which move in arms of the framework A by means of the screws H H, and to them are attached the bearings of the mandrels, the colour-boxes and the "doctors." By means of the screws H and I, and the levers K, additional pressure is given to the cylinders D, the top cylinders D are pressed against the drum by the levers K, which are attached to the arms of the framework. The two bottom cylinders D are pressed against C by the levers K in a similar manner to the upper ones. a is the cloth to be printed.

In the above machine, one engraved cylinder is capable of printing one colour only. A pattern or design, therefore, in which there are several colours, will require a separate cylinder for each; all the cylinders working in one machine. When this is the case, the pattern is cut up into as many parts as there are colours, each part being engraved upon a distinct cylinder. By arranging the respective cylinders so that each will print the colour in its right position, the vari-coloured pattern will be produced in its integrity. Upon the proper and accurate adjustment of the cylinders, and the consequent exact fitting in of the mordants or colours with each other, the distinctness of the design depends. The material surrounding the drum generally consists of a particular kind of strong, coarse woollen cloth, known as *lapping*. It is about half an inch thick. The blanket is a woollen web with the ends sewn together. It is about 40 yards in length and should be of equal thickness throughout, as well as of uniform texture and elasticity. A blanket of good quality will print 10,000 pieces; whenever saturated with mordant it is washed, and is then employed for covering the table used in block-printing. Blankets made of cotton, covered with India-rubber and of vulcanized* India rubber are sometimes employed. They are said to act satisfactorily, and to be more economical than the woollen ones, since they last longer, in addition to which they can be more easily washed. Spite of these advantages, however, the woollen blanket is largely used. Processes for dispensing with blankets have, from time to time, been introduced, but they have not met with much favour. The " colour doctor," or sharp instrument used to scrape the superfluous mordant from the engraved cylinder as it revolves, is generally made of steel. It must be so tempered as to be capable of receiving a fine edge, but not so hard as to resist being cut with a file. For mordants which act rapidly on steel, brass and nickel " doctors" are sometimes employed; neither of these metals, however, can be as accurately tempered as steel, besides which, being softer, "doctors" made of them are more quickly worn out, and are more difficult to clean. But all colours containing salts of copper, necessitate the use of brass, composition, or nickel "doctors." For if steel be used, then the nitrate, acetate, or sulphate of copper in the colour, immediately deposits its oxide in irregular layers on the sharpened edge of the doctor, and thereby permits colour to pass under it that should be kept back.

---

* An india-rubber blanket, the invention of Mr. JAMES CHADWICK, of the firm of CHADWICK & Co., of Manchester, is in occasional use. This blanket, which was patented by the inventor, consists of a thin woollen fabric known as camlet cloth, two folds or plies being held together by a thin layer of india-rubber, which is afterwards vulcanized. Owing to its thinness, and to its being made of wool, it combines pliability with durability, and being waterproof, after every passage through the printing machine, it is washed and dried by means of an apparatus situated behind the machine. By this arrangement a clean surface is secured after every registration. Mr. CHADWICK's patent has now expired.

Previous to either printing or dyeing silk, wool, cotton, or linen, each of these fibres has to pass through several cleansing processes, in order to secure the desired results.

SILK must be boiled for some hours in soap and water, until the gum natural to the silk, as it comes from the cocoon, and other matters used in its manufacture, are entirely removed. It is afterwards suspended in a sulphur chamber filled with the fumes of sulphur, for eight to twelve hours. This gives a whiteness to the fibre. If, however, the goods are for dyeing, and the whiteness is not material, the sulphuring is dispensed with. After the soaping or sulphuring, the goods are well washed and dried, when they are ready for dyeing or printing.

WOOLLEN GOODS, before being printed or dyed, are generally what is called "crabbed," in either weak ammonia or weak soda lye. The finer qualities or styles are crabbed in soap and water; whilst for common colours putrid urine, varying in temperature from 140° Fahr. (60° C.) to boiling, is used. If intended for styles where a pure white is required, the goods are also sulphured previous to washing and drying, for printing or dyeing. The sulphuring is omitted in those cases in which the colours would be injured. The all important object is to effect the removal of yolk and oil.

Before printing or dyeing COTTON or CALICO, the goods are dressed or singed to remove the fibrous down from the surface. The singeing is performed by passing them perfectly straight and even over red-hot copper or iron plates, at the rate of about 100 yards per minute; or by means of TULPIN's singeing machine, which causes the cotton or calico to be moved along the lateral part of a coal-gas flame. The flame is kept steady by means of a ventilator, which draws off the gases arising from combustion. The flame being mixed with air, the same as in the Bunsen lamp, perfect combustion is effected, consequently no smoke is given off, and the cloth is neither discoloured nor blackened.

The goods are next passed through a washing-machine, and then through milk of lime into a vessel termed a *kier*, where they are boiled from five to eight or twelve hours in the usual quantity of water, heated by high or low pressure steam; steam of low pressure being preferable, since the boiling is then more effective and occupies less time. The goods are afterwards well washed and steeped for some hours in hydrochloric acid at 1° to 2° Twaddle, after which they are again well washed and boiled in the kier in soda ash, or carbonate of soda, or resinate of soda, the proportions being, say, 140 lbs. to two tons weight of dry cloth. In this they are boiled from six, eight, to twelve hours, afterwards washed and steeped for three or four hours in a solution of chloride of lime at $\frac{1}{2}$° to 1° Twaddle. They are again washed, then steeped in hydrochloric acid at 1° Twaddle for two hours, rewashed, squeezed and dried.

LINEN, before being printed or dyed, is treated in much the same manner as calico, but is more frequently boiled in ash lye than in lime.

The pieces intended to be printed on by the cylinder machine, stitched together by machinery, or sometimes gummed, mostly in lots of forty, are wound by means of an apparatus called a *candroy* on to a wooden roller, which is fixed at the back of the printing-machine.

During the time it is being unwound from this (see *a* and *b*, fig. 3, page 65), the fabric is made to pass over wooden rollers, and is additionally stretched tightly by means of a pulley attached to the axis of the roller, so that when it arrives at the drum it is in a smooth and uncreased condition, and is kept in position by a boy placed behind the machine. This boy also removes any loose fibre from the cloth. The printer stands in front of the apparatus, and adjusts his pattern on a piece of coarse cloth, attached to the end of the nearest piece. This man also has the colour boxes under his supervision. After forty pieces have been worked off the machine is stopped, and previous to another batch being passed through it, the " doctors" are examined, and, if needful, sharpened by means of a file.

It is necessary that the pieces upon which the mordant has been printed should be immediately dried. This is done by passing them from the printing rollers, over steam chests and cylinders. If the style is for steam or topical prints, the goods are passed on to the steaming box described further on. If for madder styles, the goods must be passed through the ageing chambers, where the mordants undergo a chemical decomposition which converts them into a state the most favourable for the subsequent operations.

For instance, the mordants generally employed in this style of printing are the acetates and pyrolignites of iron, used either separately, or in combination with the salts of alumina,* and the salts of copper and iron used with catechu, and these salts not possessing the essential quality of insolubility, would fail to attach themselves either to the dye or to the cloth. When, however, the calico printed with them is exposed to the heat and moisture of the ageing-room, the acetic or pyroligneous acid is in great part driven off, and owing to the absorption of oxygen, either an insoluble oxide or hydrated peroxide, or a subsalt, is left behind, confined to the space on which the pattern was impressed, which, after being subjected to the treatment described further on, enters into insoluble union both with the cloth and subsequently with the tinctorial body. Goods were formerly aged by hanging

---

* Alumina salts alone require no ageing previous to their being dunged.

them in folds from the roof of a very spacious airy building, kept at a mean summer temperature, and by the introduction of aqueous vapour into the apartment. The hygroscopic condition was sometimes ensured by mixing the mordants with a deliquescent salt.

In the process of ageing, as now carried out, the exposure of the mordanted cloth by hanging in folds is got rid of altogether. After it has been mordanted, it is passed through what is called the "ageing machine." The ageing machine was originally introduced by Mr. THOM, of Mayfield Print Works. Shortly after, however, Mr. W. CRUM, of Thornlie Bank, by increasing its size, and giving it a more practical form, soon caused it to be extensively adopted, and it is now in almost universal use. The ageing machine consists of a chamber about 36 feet long, 20 feet high, and 13 feet wide. It is fitted at top and bottom with rollers, over which the mordanted cloth is made to pass at the rate of 60 to 80 yards a minute. The chamber, which is made of wood, and is air-tight, is fitted with steam pipes and jets, which supply it with the necessary heat and moisture. It also contains a DANIELL's hygrometer, which, when the chamber is in use, should indicate a temperature of $75°$ to $80°$ Fahr. ($23·28°$–$26·6°$ C.), and $70°$ to $76°$ of moisture. Ageing machines are manufactured by MATHER & PLATT.

After the mordanted goods have passed through the ageing chamber, they are folded in bundles of convenient size, and placed in a room, heated to about the same temperature, till next morning, when they are ready for dunging. But the tissue is not yet in a fit condition to enter the dye-beck, since, when it leaves the ageing house, there is always adhering to its printed portions a sensible quantity of superfluous and unoxidized mordant, and a small quantity of undecomposed acetate or pyrolignite of iron or alumina, as the case may be. If these were allowed to remain, they would not only cause a waste of dye-stuff, but, in combination with this, would spread over the fabric and impregnate with colour those parts intended to be kept white. In addition to the excess of mordant, the substances with which it had been thickened must also be got rid of.

Simple immersion of the fabric in warm water, and subsequent rinsing in cold, would remove these; but such a course of treatment would be inapplicable to the superfluous mordant, inasmuch as this would dissolve in the water, which would then colour the cloth, and thus, more or less, obscure the design on it.

Formerly the superfluous mordant was got rid of by passing the aged goods through a bath of cow-dung; whence this particular stage of the dyeing operation derived its present name of "dunging;" called by the French printers *bousage*. CAMILLE KŒCHLIN, a French authority, ascribed the efficiency of the dung to the presence of an albuminous constituent, which, he believed, combined with the alumina and iron of the acetates of these bases dissolved by the hot water of the bath, and formed with them insoluble precipitates, which fell to the bottom of the vessel. It is, however, more probable that the efficiency of the dung was due to its containing certain salts, such as phosphates and silicates, the acid radicles of which, seizing the bases of the free mordant, precipitated them in the form of insoluble salts, which were incapable of combining either with the fibre or the colour.

This latter view receives strong support from the fact that similar salts are now almost universally used as dung substitutes. Those in use are: 1, The double phosphate of soda and lime; 2, arseniate of soda; 3, arsenite of soda; 4, silicate of soda; 5, silicate of lime. The selection of these substances depends upon the particular style for which they are required. Phosphate of soda and lime was first patented by MACQUER; silicate of soda, by SCHLIEPER; and silicate of lime, by HIGGIN, as being an improvement upon silicate of soda, the alkalinity of which was sometimes objectionable.

On the Continent, bran, which is rich in phosphates, is used as a dung substitute. Deeper and brighter colours result from the employment of the dung substitutes than from the dung itself; in addition to which the white parts are left clearer, since the dung occasionally imparted a greenish stain to these. The varying constitution of the dung, as giving rise to uncertain effects, also constitutes an important drawback. Lastly, a great saving of labour and time is gained when the salts are employed.* The dung-bath is a large trough filled with weak solutions of the arseniate or silicate of soda, or some other salt used as a dung-substitute. In the trough are fixed two rows, about twenty in a row, of rollers, one row at the top, the other at the bottom, over and under which the mordanted cloth, sewn together by the ends, and extended to its full width, and free from folds, is drawn as expeditiously as possible, the bath being maintained at a temperature at from 160° to 180° Fahr. (71°–82·2° C.). It is important that the pieces should be made to pass through the fly-dung bath at such a rate of speed as not to remain in it more than a minute or a minute and-a-half, and also that the bath should, from time to time, have strong liquor added to it, so as to keep it up to the required strength during the time the pieces are being carried through it.

* " Experience proves that in the case of the best dung substitutes, a final turn in cow-dung before dyeing is advantageous, it being better for the mordanted oxide that it should go into the bath in a partially saturated state than in a state of the highest activity. In a majority of cases the colours will be more solid, brighter and faster when the combination between the mordant and colouring matter is slow and gradual than when it is rapid."—*Practical Handbook of Dyeing and Calico-printing,* by W. CROOKES, F.R.S.

After they leave the trough, the goods are thoroughly rinsed, and then drawn spirally through a vessel resembling a dye-beck, filled with a very weak dunging solution heated to 140° Fahr. (60° C.). After this second dunging, termed *cleansing*, and which is known in French as *déyommage*, the principal use of which is to remove the substances used to thicken the mordant, the goods being well washed in a machine made for the purpose, are in a condition to be dyed.* In the dyeing of calico printed goods, the process is performed in oblong or wooden vessels called dye-becks.

These becks are usually divided lengthwise by a perforated diaphragm, under which runs a perforated pipe for the admission of steam. Above the dye-beck is a winch or reel connected with a driving shaft. The necessary quantity of water having been run into the beck, the mordanted, aged and dunged pieces are laid in a row, in seven or eight lengths of two each, across the beck, the ends of each being tied together so as to form an endless web, which is wound round the winch, the diaphragm and steam-pipe, the greater part of the cloth being in the front part of the beck. The dye-decoction or solution of the colouring substance being then added, the driving shaft is set in motion and the steam gently turned on, and the liquid gradually raised to the boiling point, the pieces being in succession dragged through the dye-bath all the time the winch is being turned, when the tinctorial substance is kept equally diffused owing to the agitation induced by the moving cloth. Some becks are divided into compartments, an arrangement which diminishes the chance of the pieces getting too much mixed together, and the better ensures their taking the dye more

---

* The above system of dunging is also suited for the artificial alazarine dyes, the only difference being that when these colours are employed, the second dunging is to be repeated.

equally. When the goods have been sufficiently long in the beck, they are removed and thoroughly washed in a washing-machine.* After madder printed goods are dyed, dunged, and washed, they have further to be brightened and cleansed. The madder colours, as they come from the dye-beck, are dull, whilst the whites or unmordanted parts, even after being thoroughly washed, are tinted with the red colouring matter of the madder. It is therefore necessary both for the brightening of the colours and the clearing of the whites, to subject the goods to repeated soapings at temperatures varying from 140° and 180° Fahr. (60°–82·2° C.). The soapings are performed in becks, which are of similar construction to dye-becks.

In order to obtain bright madder reds, scarlets, and pinks, it is further necessary, after the goods have been dyed and washed, and previous to their being soaped, to have them dried, and afterwards padded through an emulsion of oleine (saponified castor oil), consisting of 1 part of oleine to 16 parts of water. They should then be dried and hung in a steaming-box, with the steam at a pressure of from one to three pounds to the square inch. They should next be treated according to the directions already given for brightening and cleansing.

Sometimes when bluer pinks are required, weak tin is

---

* This operation—which has for its object the removal of the superfluous and uncombined dye—is more particularly in piece-dyeing—conducted as follows:—The pieces are first vigorously agitated in a large reservoir of water, from whence they pass between two cylinders, being subjected in doing so to a heavy downpour of water; from the cylinders they descend into a smaller reservoir supplied with water from a wide pipe, and finally they are carried through a second pair of cylinders, being exposed during their passage to a powerful fall of water. Improved washing-machines are supplied by the following amongst other makers:—MATHER & PLATT; FURNIVAL; and BARLOW.

used between the soapings. Sometimes, in order to obtain the necessary purity of colour in the ground, or the whites, it is necessary to pass the goods through a weak solution of chloride of lime. After calico goods are printed, they generally require to be subjected to the process of "finishing," before they are in a condition to be delivered to the merchant for sale.

To the outsider, the operation of finishing may seem a matter of no very great importance, but by the calico printer it is regarded in a very different light. When making his purchases of printed goods from the manufacturer, the buyer generally requires to know their length. width, weight, &c. Now, before the goods are printed, all foreign matters that have been added to the cotton, for the purposes either of manufacture or gain, must be removed.

In some cases, removal of these extraneous bodies amounts to a loss of as much as 10 per cent. Besides this, the goods having been made to pass through so many operations, become soft, loose, and rough, in which state they are very unsuited for the market. To remedy this, they are passed through a machine called a "mangle," a vessel charged with starch water varying in consistence according to the nature of the finishing required, and then immediately after over drying cylinders. Finally, they are passed once or twice through a calender to give them the necessary smoothness. Owing to the varied requirements of the different markets, and to the numberless uses to which printed calicoes are put, there is considerable diversity in the manner in which finishing is carried out. Hence we have "embossing finish," "glazed or Swissed finish," "bath finish," &c. &c.

MORDANTS.

As already explained, mordants are substances employed, when necessary, by the dyer and calico-printer, to fix his colours to the fleece, yarn, or tissue. They are in extensive use, because, with the exception of indigo, safflower, annatto, and the aniline colours, there are scarcely any tinctorial substances the essential principles of which can be fixed to textile fabrics without their intervention.

In its extended meaning, the word "mordant" signifies any agent that accomplishes this fixation, not only by entering into insoluble union both with the colouring principle and the textile fibre, but that also acts as an intermediary without itself entering into combination. The salts of alumina, iron, tin, and many other compounds belong to the first class; whilst as illustrations of the second may be taken oxygen, which by its action on the reduced indigo of the blue vat, causes the adhesion of the resulting indigotin to the fabric; and the acid employed to neutralize the alkaline lye holding in solution the colouring principles of safflower and annatto, which, upon withdrawal of the alkali, are precipitated upon the cotton yarn or cloth, and enter into indissoluble union with it.

The term "mordant," however, is generally applied to solid bodies, and to solutions of these.

The following list includes most of the mordants in use :—
Salts of :—

Aluminium.	Potassium.
Antimony.	Sodium.
Arsenic.	Tin.
Chromium.	Also :—
Copper.	Albumen(from egg and blood).*
Iron.	Certain fatty oils.
Lead.	Casein (lactarine).
Manganese.	Gelatin.
Mercury.	Tannic acid.

---

* Albumen, casein, gluten, and other coagulums, which were at one

One or more of the above metals form salts with one or more of the following acids : arsenic, boracic, hydrochloric, hyposulphurous,* nitric, phosphoric and silicic. The oxides of the metallic or inorganic compounds in the above list exhibit a much stronger affinity for the fibre and the colouring principle, and enter into a correspondingly more intimate union with it than the inorganic substances.

The oil mordants first used in Turkey-red dyeing are supposed to owe their efficacy as mordants to exposure to the air, and consequent absorption of oxygen, whereby they become converted into insoluble bodies capable of combining with certain colouring matters and fixing them on vegetable fibres. The action of bitartrate of potash or cream of tartar as a mordant has yet to be defined.

The essential conditions of a mordant are, that it should be in a soluble state when applied to the fabric, and should become fixed and insoluble when combined with it. Unless it be in the liquid form it cannot penetrate and thoroughly and equally impregnate the cloth, and if, after its absorption, it were not permanently fixed, it would be washed out by

---

time largely used as mordants or agglutinants for fixing the coal-tar colours, have, owing to improvements in the manufacture of these, been supplanted by the ordinary metallic mordants. Albumen and casein—more particularly albumen—are, however, very extensively employed in fixing to textile fabrics insoluble powders and pigments, such as ultramarine, chrome green, vermilion, chromate of lead, &c. These pigments are merely fixed mechanically or by surface adhesion, owing to the coagulation, mostly by means of heat, of the albumen or casein.

* KOPP first suggested the use of hyposulphite of soda as a mordant. He states that it is entirely soluble in water, and that in decomposing, it forms no product that is destructive of the fibre. It is also cheap, gives rise to full-bodied colours, is more rapidly and completely fixed than alumina, and more effectually prevents fixation of iron. Gum, British gum, or torrefied starch, may be used as thickeners.

the liquid that in the first instance had been used as its menstruum.

Another condition is that it should possess a strong attraction for the colouring matter, and for the cloth also, and form with them a compound of such stability as to be irremovable when exposed to the ordinary agencies of friction, soap, water, light and air.

The affinity of the mordant for both fibre and colouring principle should be nicely adjusted; for if the attraction it possesses for these be excessive, the resulting dye will be deposited in an uneven and spotty manner; on the contrary, if it be too weak, or some other substance is present which exercises a more powerful attraction, the dyed goods will be poor, and thin in shade, and the colour will probably be fugitive.

Again, the use of a mordant having a greater affinity for the colouring matter than for the cloth, should be particularly avoided, or the mordant combining with the greater part of the tinctorial substance, and forming with it an insoluble lake, which falls to the bottom of the dye-pan, will, of course, more or less prevent its deposition upon the fabric. Those bodies, the components of which are held together by a loose affinity, make the best mordants.

The fixing of the metallic mordant is obtained by various methods, sometimes the salt is converted into the insoluble base by volatilization of its acid. Thus, when textile goods have been mordanted with the acetates of alumina or iron, and are afterwards "aged," the acetic acid escapes, leaving behind, according to some chemists, an insoluble hydrated oxide of aluminium or iron, according to others a basic body attached to the cloth. When pernitrate of iron is employed as the mordant, the acid, which is only feebly combined with the base, is removed by means of dilution with water, whilst the peroxide of iron is precipitated upon and enters into union with the fibre. Again, when a fabric is dipped into a bath of aluminate of sodium, and afterwards passed through a solution of chloride of ammonium, the resulting insoluble oxide of aluminium is deposited upon it; or the same thing occurs by simple exposure of the fabric, after its removal from the aluminate of sodium solution, to the air. when the atmospheric carbonic acid combines with the soda, and leaves the alumina combined with the cloth. Again, when cotton goods, which are to be dyed with the aniline colours, are first mordanted by passing through a solution of silicate of sodium, and afterwards through a dilute acid, the liberated silica becomes fixed on the cotton.

The function of the mordant is not confined to the fixation of the colour alone, since in the majority of cases it increases in a marked degree, the brilliancy and depth of shade of the dyed fabric. Thus alumina very greatly increases the intensity of the colouring principle of madder; and perchloride of tin, when added to the cochineal dye-beck, gives rise to beautiful scarlet and crimson shades that would not be brought out by water alone. It must not be omitted to be mentioned also, that mordants very frequently render the colour faster. The different shades of colour in fabrics dyed with the same mordant, depend upon the degree of concentration of the mordant. Madder, with a strong iron mordant, gives a dark purple, with a weak one a lilac. With a strong alumina mordant, this dye-stuff gives a deep red, and with the same mordant diluted, a pink shade. A mixture of the acetates of alumina and iron, according to its strength, gives rise to varying shades of chocolate. The same mordant gives rise to different colours with different dye stuffs. Alumina which with madder produces pinks and reds, with logwood gives rise to greyish purples, and with old fustic, to yellows.

The nature of the fibre, as well as the mechanical condition of the surface of the fabric, exercise a considerable influence on the activity or otherwise of the mordant.

In dyeing woollen fibres, fast colours are obtained with tin mordants; whereas this latter acts very feebly when used with cotton goods. On the contrary, iron acts as a very powerful mordant when applied to cottons, but is very difficult of application to woollen goods.

In the choice of mordants, those of a strongly acid or alkaline nature should be avoided, since they corrode the fibre, and by thus rendering it incapable of properly absorbing the dye, give rise to meagre, impoverished, flattened and lustreless colours. Acid mordants are best adapted for wool and worsted goods; neutral ones for silk; whilst in cotton dyeing and printing, excess of free acid is decidedly objectionable.

The mordant is usually applied to cotton before dyeing, and to wool generally along with the colouring matter, and occasionally after it has been dyed. Several mordants, after a time, suffer spontaneous decomposition. The agents which hasten this are artificial heat, exposure to light, and sometimes to a low temperature.

As strong mordants are particularly prone to decomposition, too much of them should not be prepared at a time.

Mordants that do not alter the colour of the fibre are, when practicable, to be preferred.

This is why alumina, tin and other bases destitute of colour mostly make the best mordants.

The mordants used in dyeing textile fibres of a uniform colour throughout are in the form of liquids; whereas those employed in calico-printing are made into pastes. The aim of the printer being to impress coloured figures upon the fabric, he can only accomplish this by confining his dyes within certain limits, and this would be impossible were the mordant in a fluid state, and as such allowed to spread over

G

the face of the tissue.* Hence, in order to keep it attached to those parts forming the pattern or design, he is compelled to have it in an inspissated and coherent condition.

The following is a list of the principal thickening agents, or, as they are termed, "thickeners":—

    Albumen.　　　　　　　　Gum tragacanth.
    Casein, or lactarine.　　　Molasses.
    Clay, China.　　　　　　　Lead sulphate.
    Clay, pipe.　　　　　　　　Potato starch.
    Dextrin in its different　Salep.
      forms of British gum,　Shellac dissolved in
      calcined starch and　　　borax.
      leiocome.　　　　　　　　Sugar.
    Glue.　　　　　　　　　　Wheat-flour.
    Gluten.　　　　　　　　　　,,　　starch.
    Glycerin.　　　　　　　　Zinc chloride.
    Gum Senegal.　　　　　　Zinc nitrate.

Of these, some are only used for particular styles or colours, whilst China-clay, chloride of zinc, and sulphate of lead, are more frequently used as *resists* or *reserves*, than as thickeners. The above substances vary greatly in thickening power. Thus 11 oz. of tragacanth gum are equal to 20 oz. of starch, 22 oz. of flour, and 8 lbs. or 9 lbs. of calcined starch.

Thicker mordants are required for cylinder than for block printing.

Any little hard particles in the mordant would injuriously affect the engraved copper cylinders. It is for this reason that China-clay and pipe-clay, previous to being used in cylinder printing, should be freed from all coarse particles by sifting and elutriation. These two substances enter into the composition of resists. Pipe-clay, combined with torrified starch, is used as a thickener in block printing. Pipe-clay is also often employed with gum as a thickener, whereby the quantity of gum required is much lessened.

Different mordants require different thickeners, and in combining the two, it is desirable to avoid the admixture of substances which decompose each other. Thus starch is ill adapted as a thickening for a strongly acid mordant, since the acid deprives the starch of its consistency. As heat assists the reaction that brings about this result, by converting a portion of the starch into glucose, the acid or acid salt should not be added to the starch until this is cold. Acid mordants, however, do not affect British gum or gum Senegal in the same manner.

Some mordants, which have been thickened by starch, are liable to become fluid after a few days, and are therefore open to the serious objection of spreading over the fabric during printing. The French printers are said to remedy this by adding ½ oz. of spirits of wine to every quart of mordant.

Mordants that are used in the production of red and pink colours are usually thickened with starch. When starch is used, it is necessary to boil the thickened mordant containing it, keeping it, at the same time, well stirred till cold. The mordant so prepared must not be used hot. For iron mordants starch makes a better thickener than British gum.

It is important to remember that starch and wheat flour, and all their derivatives, such as dextrin, British gum, &c., are reducing agents, and thereby prevent oxidation of the colours. Mordants thickened with starch give deeper tints than gum. The natural gums should not be used as thickeners for mordants containing basic or sub-acetate of lead, basic alum, solutions of tin, or nitrate of copper or iron, since all these substances coagulate the gum. Starch is occasionally adulterated with gypsum, sulphate of barium and chalk. Mordants containing dextrin are liable to spoil sooner than when made with gum.

Whenever its use is practicable, gum is preferable to starch as a thickener, since it imparts greater transparency to the dye, does not alter nor tarnish the colours, nor weaken the mordant, whilst the unfixed portion is more easily removable by washing from the fibre. But against these advantages are to be set the tendency of the gum to dry too speedily, and to form an impermeable coating on the cloth, which gives rise to poor and feeble tints.

The best kind of gum only should be chosen for dyeing pink-reds and rose-reds. Except with pipe clay, the gums are not very miscible with the other substances used for thickening. Solutions of the natural gums quickly become acid.

Gum Senegal is chiefly used in roller printing. Gums vary in the degree of viscosity they impart to water, or, in other words, in thickening power. The usual way of determining the comparative viscosity of two or more samples of a gum, is to make a solution of each, under precisely similar conditions, and to pour the solutions one at a time through a glass funnel, with the tube drawn out to a fine point. The length of time it takes each solution to run through the funnel is then carefully noted, and the results compared; that solution of course being the strongest which has been the longest in passing through the funnel. The natural and artificial gums form the best thickeners for printing purposes. When used with the acetates of alumina and iron, they give paler colours than torrified starch. A much smaller quantity of thickener is required when it is employed with salts that have a coagulative power. Colours prepared with casein are liable to decompose. Solutions of albumen also decompose very quickly. They are best preserved by adding to them about 1 per cent. of arsenite of soda, or arsenious acid. Oil of turpentine is also used for the same purpose. A very small quantity only is necessary.

It may be stated that the larger the quantity of thickening used, the lighter will, in most cases, be the shade of colour; and that the most lustrous tints are obtained, when, by judicious thickening, the mordant is kept as much as possible to the surface of the cloth. If the latter precaution be observed, a great saving of dye is also gained. A mordant may be quickly rendered thicker by the addition of starch or farina, without any detriment to the resulting colour.

A special apartment is reserved for the preparation and mixing of the mordants* and their thickenings. The "colour-room," as it is called, is usually a large well-ventilated building, with a range of colour pans placed sometimes at one end, and sometimes on one side of it. The annexed drawing exhibits a series of mordant-pans manufactured by STOREY, of Manchester:—

FIG. 4.

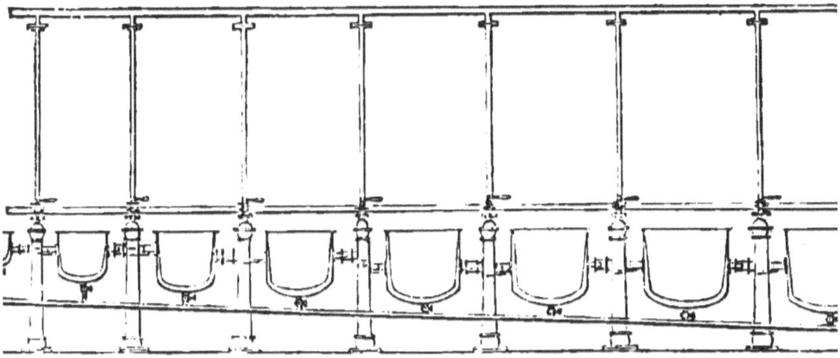

The series consists of eight jacketted copper pans, varying from 1 to 28 gallons in capacity. The pans are supported on cast iron pillars, and are so arranged, that when the mordant requires to be emptied out, they are turned over by

---

* In many works on dyeing and calico-printing, mordants are very frequently spoken of as "colours," a looseness of phraseology which is occasionally likely to lead to their being confounded with the dye.

means of a brass stuffing box, attached to each pan, and to each pillar. By means of the copper pipe A, steam can be admitted down the pillars, as far as the stuffing boxes, and thus into the jacketted space surrounding each pillar; D, is a condensing pipe for carrying off waste steam and water; C, is a copper pipe for letting cold water into the casing of any pan, and cooling its contents. The mordanting substance and the thickening being put into the pan, are boiled until perfectly smooth, being all the time constantly stirred by hand or machinery. The steam is then shut off and cold water admitted into the jacketted space, so that the contents of the pan are cooled. But before the mixture is in a fit state for use, it must be strained in order to keep back any gritty particles which would scratch the engraved portions of the copper cylinders. The straining is very frequently performed by means of an apparatus invented by DOLFUS & MEIG.

STYLES OF CALICO PRINTING.

In most of the works on calico-printing, the descriptions of the operations adopted to produce particular effects, are collected under separate heads or divisions, called "Styles."

The following list includes all these styles:—

1. Madder and Alizarin Dyed . . style.
1a. ,,   ,,   ,,   ,, . . blocked.
2. Reserve . . . . . style.
3. Garancine . . . . . ,,
4. Padding . . . . . ,,
5. Indigo . . . . . ,,
6. China Blue . . . . . ,,
7. Indigo Discharge . . . . ,,
7a. Turkey Red Discharge . . ,,

8. Prepared Steam . . . . style.
8a. Unprepared „ . . . . „
9. Spirit . . . . . „
10. Bronze . . . . . „
11. Pigment . . . . . „
12. Extract or Topical Fast . . „

In the earlier and more primitive days of the art of calico-printing, this classification may have been useful; but owing to the rapid increase in the number of new methods, and to the practice of sometimes placing these under one of the old designations, as well as to the variety of effects that are obtained by combinations of these methods, the above list is eminently inadequate and unsatisfactory, and wanting in the great merit of particularity.

1. **Madder and Alizarin Dyed Style.** *Syn.* DYEING UPON MORDANT.—The simplest form of this style of calico-printing consists in the production upon a white ground, by the aid of one or two mordants, of patterns of one, two, or more colours, or shades of colour, the different depths of tint being produced by varying the strength of the mordant.

Another modification of this style, whereby very great diversity of effect is gained, is that in which "resists" and "discharges" are employed. Thus, suppose it is desired to introduce into a fabric, printed over with a particular design in purple or light chocolate, for instance, some patterns in red or white. It is only necessary to first print these patterns on the cloth, the red with a special kind of mordant known as "resist red," and the other with thickened lime-juice, and then to print it over in iron or a mixture of iron and alumina mordant with the predominant pattern, and then after dyeing and dunging, to put the piece into the madder beck, when the purple or chocolate design will cover every part of the cloth, save that protected by the "resist" mordant and the lime-juice, the latter of which dissolves and leaves a clear white, and the former a red pattern.

Another variety of the madder colour style is that known as the "French Pink Style," by which designs are produced in red and pink. If more than one shade of colour is required, the calico is printed with a red or alumina mordant of varying degrees of strength, the whites being obtained by means of lemon-juice. The whole is then printed over with a mordant, which will give a very pale red, and then run into the madder-beck.

The dyed goods obtained by the above processes are fast in colour, and wash well.

In the above processes, madder-root has been almost, if not entirely, superseded by artificial alizarin and anthrapurpurin.

a. Madder and Alizarin Dyed Style, Blocked.—This process of printing is performed by hand, after the goods are dyed. By this method certain blue, green, violet, yellow, and other bright colours that cannot undergo the operations of dunging, dyeing, and soaping, can be applied to the print and add greatly to its brilliancy.

2. **Reserve Style.** *Syn.* Resist Style.—This is a modification of the first or madder style, or rather a union of this with the "steam colour style." The fabric is first printed with acid resists of lemon-juice and caustic soda, or some suitable substance; then mordanted with red or iron liquor, or a mixture of both, according to the required colour, and afterwards aged, dunged, dyed in the madder bath, and cleared; the parts which are left white are afterwards blocked in with pigment colours, and steamed.

Sometimes what is termed a "resist" paste is blocked in over the whole or a portion of the design originally produced, and a small coloured pattern is printed over all. The reserve style may be worked in numberless ways, and so give rise to a great variety of effects.*

3. **Garancine Style**. This is carried out in a manner very similar to that of the madder style, except that the temperature of the beck is lower than that used in madder dyeing. Garancine goods require much less soaping than madder ones. With the exception of the purples, the colours given by garancine are greatly inferior in brilliancy to those yielded by madder root; in addition to which they are not so fast. The most predominant hues are dark reds, browns, and oranges. In garancine dyeing, catechu is largely used as an accessory.

4. **Padding Style**. *Syn.* PLAQUAGE STYLE.—This is also a variety of the madder style. The fabric, after being padded all over with the requisite mordant, is dried in the padding flue. A design is next printed on it in acid discharge (mostly a compound of lime-juice and bisulphate of potash thickened), with the result that after the cloth has been aged, dunged, dyed in the madder bath, and cleared, white patterns appear upon a ground of uniform colour. These white patterns or spaces may be afterwards printed upon in steam or pigment colours.

5. **Indigo Blue Style**.—A pattern is first printed upon the goods with a "reserve" paste in those parts which it is

---

* Solution of citric acid, or lemon-juice at 15° Twaddle, thickened with torrified starch, forms the best resist paste for iron and red liquor mordants; when used for Garancine Styles the lemon-juice should be at 30° Twaddle. Oxalic and tartaric acid, as well as bisulphate of potash, are also recommended; but they are not so satisfactory as lemon-juice or citric acid.

We append formulæ for a protecting resist paste for chintzes:—Water, 6 gallons; neutral arseniate of potash, 15 lbs.; pipe-clay, 40 lbs.; calcined farina, 30 lbs. The protecting resists are usually applied by the block.

intended shall be white. The goods are then dyed to the desired shade in the indigo vat (see page 142), when the indigo is deposited on every part of the cloth, except where the "reserve" has been applied. After passing the goods through a dilute acid bath there is obtained, on a blue ground, a number of white patterns, which may be blocked with steam colours. Or yellow or orange designs may be produced by combining with the reserve paste a salt of lead and then passing the fabric through a weak acid solution.

6. **China Blue Style.**—This is the reverse of the previous style, since a blue design is printed on a white ground. It derives its name from the resemblance of the pale blue colour of the design to certain kinds of porcelain or china. The process is conducted as follows:—The goods are printed with a mixture composed of finely powdered indigo and acetate of iron. They are then passed through six successive indigo vats. Of these, the first two contain lime; the third, sulphate of iron; the fourth, caustic soda in solution; the fifth, a dilute solution of sulphuric acid; and the sixth, water. When the desired tint has been reached, the goods are washed a second time, passed through dilute sulphuric acid, and finally washed again.

7. **Discharge Style.** *Syn*. TURKEY RED WITH DISCHARGES. ENLEVAGE.—Discharges are mixtures which, when printed upon cloth previously dyed of some uniform colour, remove the colour from the printed parts, thereby leaving a white design upon a coloured ground. The term "discharge style" is more particularly restricted to the process by which white or coloured designs are obtained upon a Turkey red or indigo ground. The operation is performed as follows:—After the discharge mixture, in the form of a highly acid mordant, has been printed upon the dyed and dried fabric, this is next passed through a solution of bleaching powder, which removes the colour only from those parts where the discharge has been

applied.* If to the discharge there is added a salt of lead, and the piece is afterwards passed through a solution of bichromate of potash, the pattern will be yellow on a red ground. The same discharge effects are produced equally well on goods dyed red and pink, when the white surface of the fibre is covered. By proper methods different coloured designs may be thus obtained.

8. **Steam Colour Style.** (Prepared).—This process is very largely employed for fixing on woollen and worsted goods, mixed fibres such as delaines and cobourgs, and very frequently silks. It is also the usual method adopted for fixing the aniline colours upon cotton. The goods are first worked in a bath of stannate of soda, and then passed through a weak solution of dilute sulphuric acid, technically known as "sours." After being drained in the centrifugal

---

* The above is the process of Kœchlin & Thompson. To guard against the contingency of the discharge spreading beyond the desired limits, a considerable excess of lime or chalk is added to the bleaching powder. Menteith's method consists in placing the dyed and properly smoothed fabric in folds, and tightly pressing them between two leaden plates, perforated with the intended pattern. A solution of bleaching powder, very slightly acidified with hydrochloric acid, is then poured into the top perforations and makes its exit at the bottom ones, removing the colour only from those parts of the fabric not protected by the lead.

The following are said to be useful formulæ for discharges :—

1. White discharges on Turkey red.—Water, 1 gallon; tartaric acid, 10 lbs. ; pipe-clay or china clay, 7½ lbs. ; gum water, 1 pint.

2. Bichloride of tin, 1½ lbs. ; hot water, 1 gallon; tartaric acid, 9 lbs. ; pipe-clay, 10 lbs. ; gum Senegal water, 3 quarts. The above can only be properly applied by blocking.

3. Discharge for machine work for light designs :—Water, 1 gallon; tartaric acid, 6 lbs. ; starch, 1½ lbs.

4. Discharge for red and production of blue on the discharged spot :—Muriate of tin, 2 gallons; Prussian blue pulp, 1 gallon; water, 1 gallon; tartaric acid, 5 lbs. Dissolve and mix, and then add 2 gallons of thick tragacanth gum water.

machine and dried, they are printed with a mixture consisting of the required colouring matters and their mordants; after which they are steamed in an apparatus called the "steam chest."

The steam chest is made of iron, and is usually about 9 feet high, 6 feet wide, and 12 feet long. At one end are accurately fitting doors, which, when necessary, can be tightly closed, and kept in position by means of bars and screws. The chest is fitted with a double bottom, the upper portion of which is on a level with the floor of the room, and is perforated with a number of holes, which admit and distribute steam escaping from a perforated pipe fixed under this false bottom, and extending round three sides of the chest. Upon the false bottom and parallel with the sides of the chest, a tramway, which is carried into the room, is fixed, and upon this tramway, there runs a carriage consisting of wooden rods fixed in an oblong frame, for holding the goods to be steamed. Much experience and care are required in adapting the temperature and degree of moisture to the different description of goods, colours and mordants.

8*a*. **Steam Style.** (UNPREPARED).—A variety of this style is in use for printing goods intended for the Eastern markets, which goods are known as "unprepared steams." The cloth is simply bleached and afterwards printed with aniline colours thickened. In this style mordants are sometimes dispensed with, and the goods are not even steamed. The colours are fugitive when exposed to the light, and easily removable by water.

9. **Spirit Colour Style.**—The colours produced by this style are very like steam colours, but they are not so durable. A large number of acids (chiefly chlorides of tin, technically called "spirits") and metallic salts enter into the composition of the mordants. The goods are not steamed

after being printed, but are merely very carefully dried after some hours exposure to the air, and then washed.

10. **Bronze Colour Style.** *Syn.* MANGANESE BRONZE STYLE. This style is now little used. The goods are padded through a solution of protochloride of manganese, dried, afterwards passed through a solution of caustic soda, washed and then the manganese is converted into peroxide by working for some minutes in a solution of chloride of lime.

11. **Pigment Colour Style.**—Pigment colour printing consists in attaching insoluble colours such as those used by the artist, to the fibre, by the aid of albumen, and fixing them by means of steam.

12. **Extract or Topical Fast Style.**—The increased adoption of this style, which is of recent date, is largely due to the introduction of artificial alizarin. It mainly consists in printing upon cloth, which has been previously prepared with oleine or saponified castor oil, extract of madder or artificial alizarin, mixed with alumina or iron mordants, thickened along with or without, pigments or aniline colours. The goods are afterwards steamed, washed, and soaped similar to madder.

As in the production of madder work, a great number and variety of processes as well as of agents tinctorial and otherwise are employed, it follows that the greater the care and accuracy with which the several operations are carried out, and the purer the dye-stuffs and chemicals used, the more successful will be the results. The necessity of cleanliness likewise cannot be too emphatically enjoined.

We append formulæ for the different styles:—

1. **Madder and Alizarin Styles** (1, 1*a*, and 2).—In the madder as in the other styles, a great deal depends upon the observance of several little conditions, if it be wished to obtain satisfactory effects. For instance, if the calico-printer desire pure reds and pinks, his mordants must not

contain a trace of iron. The fabrics must also be free from oil or grease stains, as these dye up stains.

Again, if the print required be a light pink one, this necessitates the use of a diluted mordant. Under these circumstances a very gentle heat must be employed to dry the fabric after it has been printed, for if a weak aluminous mordant is brought into contact with a heated metallic surface, such as iron, tin or copper, its power to attract the colouring matter is diminished by at least 30 per cent.

This sensitiveness of diluted aluminous mordants when brought into contact with heated surfaces is a source of great trouble, and frequently of loss to the calico-printer; for in order that the perfect register of the design as given by the impression of the engraved roller should be maintained, it is necessary that the piece be dried within as few moments after printing as possible. The cause of this change wrought by heat in the condition of the mordant has given rise to many speculations. Some authorities affirm that the alumina, being so quickly deprived of its solvent, becomes in substance physically like horn, and is therefore incapable of gradually absorbing the colouring matter, as it does when in a less compact or solid state. Others hold the opinion that when the alumina mordant is slowly dried by heated air, the particles of alumina are absorbed by the fabric in a highly divided state, and thus offer a greater surface of attraction to the colouring principle. On the other hand, when the mordant comes into contact with a heated substance, its particles are absorbed in groups, and consequently do not present the same amount of surface, as when they are in a more minute state of division.

It will now be seen why diluted mordants intended for madder or alizarin pinks, should be dried by heated air.

In the use of diluted aluminous mordants it is also important that the temperature of the fly dung-bath, as well as the dye-beck, for the first forty minutes should not be too high.

Iron mordants do not suffer so much from over-drying as aluminous ones. Over-oxidation is most to be avoided in them, for if the iron be too highly oxidized, the lighter colours will be greatly deficient in brilliancy, and the blacks will not dye so easily.

MADDERS (Fast Work).

1. Purple Fixing Liquor.

    Water . . . . 7½ gallons.
    Acetic acid . . . . 1½ ,,
    Sal ammoniac . . . . . . 9 lbs.
    Arsenious acid . . . . . . 9 ,,

Boil until the arsenic is dissolved, and let the solution stand till clear, then decant.

2. Purple Assistant Liquor (Messrs. Mercer & Baines' Patent).

    Potato starch . . . . 100 lbs.
    Water . . . . 37½ gallons.
    Nitric acid, at 60° Twaddle . . 123 ,,
    Black oxide of manganese . . 4 oz.

When the reaction is over and the nitric acid destroyed, add pyroligneous acid, 50 lbs.

3. Black. *For Machine Work.**

    Iron liquor, at 24° Twaddle . . 4 gallons.
    Pyroligneous acid . . . . . 4 ,,
    Water . . . . 4 ,,
    Flour . . . . 24 lbs.

---

* Aniline blacks are largely displacing madder in this style.

The flour must be worked up with a small quantity of the mixed liquid, until a perfectly smooth thin paste is obtained; the remainder of the liquid is then added, and the whole boiled; lastly, 1 pint of Gallipoli oil is added. The quantity of flour must be increased or decreased to suit the strength of the engraving.

4. ANILINE BLACK.

Standard No. 1.

Chlorate of soda	$14\frac{1}{2}$ lbs.
Water	9 gallons.
Starch	16 lbs.
Dark British gum	16 ,,

Boil and cool.

Standard No. 2.

Water	8 gallons.
Dark British gum	16 lbs.
Starch	16 ,,

Boil, and add

Vanadiate of ammonia, that has previously been dissolved in 1 gill of hot water . . . . $3\frac{1}{2}$ drachms.

And cool.

Colour.

Of Standard No. 2, cold	5 gallons.
,, ,, 1, ,,	3 ,,

Mix well and strain.

5. MADDER PURPLE (dark).

Water	10 gallons.
Iron liquor, at 24° Twaddle	$1\frac{1}{2}$,,
Purple fixing liquor (No. 1)	2 pints.
Fine flour	$15\frac{1}{2}$ lbs.

Thoroughly incorporate and boil as madder black.

6. MADDER PURPLE (medium).
    Water . . . . 14 gallons.
    Iron liquor, at 24° Twaddle . . 1¼ „
    Purple fixing liquor (No. 1) . . 12 pints.
    Fine flour . . . . 18 lbs.
        Treat as dark madder purple.

7. MADDER PURPLE (light).
    Gum water (farina dark) . . . 9 gallons.
    Iron liquor, 24° Twaddle . . . 1 quart.
    Purple fixing liquor . . . . 4 pints.

8. STANDARD BROWN.
    Water . . . . 50 gallons.
    Catechu . . . . 200 lbs.
  Boil for six hours; then add
    Acetic acid . . . . 4½ gallons.
    Water to make up . . . . . 50 „
Let the whole stand for two days and decant; after heating this to 130° Fahr. (54·4° Cent.) add
    Sal ammoniac . . . . 96 lbs.
Dissolve, and let it stand for forty-eight hours. Decant, and to the clear liquid, add gum Senegal in the proportion of 4 lbs. to the gallon.

9. BROWN. *For Machine Work.*
    Standard Brown (No. 8) . . . 4 gallons.
    Acetate of copper* . . . . . ½ „

---

* The acetate of copper is prepared as follows:—
    Sulphate of copper . . . . 4 lbs.
    Lead acetate . . . . 4 lbs.
    Hot water . . . . 1 gallon.
Dissolve, let settle, and dilute the clear liquid to 16° Twaddle with water.

Acetic acid . . . . 2 quarts.
Gum Senegal water (4 lbs. to the
    gallon) . . . . 2 ,,

10. **Medium Brown.**

No. 9 . . . . 2 gallons.
Gum water . . . . 2 ,,
Acetate of copper . . . . . 1 gill.

11. **Madder Brown to Resist Heavy Covers of Purple.**

Catechu . . . . $\frac{1}{2}$ lb.
Sal ammoniac . . . . . . $\frac{1}{4}$ lb.
Lime-juice at 8° Twaddle . . . 1 quart.
Nitrate of copper at 80° Twaddle $2\frac{1}{2}$ oz.
Acetate of copper . . . . . $1\frac{1}{2}$,,
Gum Senegal . . . . 1 lb.

12. **Chocolate.** *For Machine Work* (dark).

Iron liquor at 12° Twaddle . . 3 gallons.
Red ,, at 12° Twaddle . . 9 ,,
Flour . . . . 24 lbs.
Oil . . . . 1 pint.
          Mix.

13. **Chocolate** (redder).

Iron liquor at 12° Twaddle . . 1 gallon.
Red at 12° Twaddle . . . . 6 ,,
Flour . . . . 14 lbs.
Oil . . . . 1 gill.
          Mix.

14. **Drab.** *For Machine Work.*

Standard No. 8 . . . . . . 4 gallons.
Protochloride of iron at 84°
    Twaddle . . . . 2 pints.

Solution of acetate of copper . 2½ pints.
Gum water substitute (4 lbs. to
 the gallon) . . . . 1 gallon.

15. FARINA GUM WATER (dark).

Water . . . . 5 gallons.
Dark coloured farina . . . . 30 lbs.
 Boil well together.

16. FAWNS. These may be produced by reducing the drab of formula 14 with gum water to suit shade required.

17. No. 20 PADDING PURPLE.

Water . . . . 18 gallons.
Purple fixing liquor . . . . 2 ,,
Iron liquor at 24° Twaddle . . 1 ,,
Logwood liquor at 4° Twaddle . ½ ,,
Flour (boiled as usual) . . . 24 lbs.

P.S.—If a gum colour is required instead of a paste one, as the above is, then take 18 gallons of gum water (dark farina), instead of the water, and keep out the flour. Of course avoid boiling.

18. DARK RED. *For Machine Work.*

Red liquor,* 9° to 10° Twaddle,
 but not to exceed 12° Twaddle 6 gallons.
Flour . . . . 12 lbs.

---

* The red liquor of commerce which is always used for dark reds (but not for pinks) is generally made by the manufacturing chemist, from crude acetate of lime and alum cake, and afterwards freed from iron with ferrocyanide of potassium.

19. **Standard Red Liquor.**
    Alum . . . . 20 lbs.
    White sugar of lead . . . . 12½ ,,
    Boiling water . . . . . . 5 gallons.
    Stir until dissolved, let the mixture settle, and decant.

20. **Pale Red.** *For Machine Work.*
    Standard red liquor, No. 19 . 1 gallon.
    Gum substitute water (30 lbs. gum substitute in 10 gallons of water) . . . . 3 gallons.

21. **Red Resist** (dark). *For Machine Work.*
    Resist red liquor at 18° Twaddle 12 gallons.
    Flour . . . . 24 lbs.
    Boil well, and when nearly cold, add
    Tin crystals . . . . . . 12 ,,
    Used as a resist for a chocolate colour.

22. **Resist Red Liquor.**
    Acetate of lime liquor at 24° Twaddle . . . . 90 gallons.
    Sulphate of alumina . . . . 272 lbs.
    Ground chalk . . . . 34 ,,

22a. **Resist Red Liquor.**
    Water . . . . 1 gallon.
    Alum . . . . 5 lbs.
    Acetate of lead . . . . . . 2½ ,,
    Soda crystals . . . . ¼ ,,

23. **Resist** (dark red). *For Machine Work.*
    Resist red liquor at 18° Twaddle 6 gallons
    Flour . . . . 12 lbs.
    Treat as in No. 21, and add
    Tin crystals . . . . 3 ,,
    This is used as a resist for a purple colour.

MORDANTS, DISCHARGES, RESERVES, COVERS, &c.

1. ALKALINE RED MORDANT.
    Potash Alum . . . . . . 10 lbs.
    Boiling Water . . . . . . 5 gallons.
 Dissolve and add
    Soda lye at 70° Twaddle. . . $\frac{3}{4}$,,

Make up to 12 gallons with cold water, let the precipitate settle, then wash it by decantation till the washings are tasteless. Filter, remove the precipitate from the filter, and dissolve it in 5 pints of soda lye at 70° Twaddle. Boil, make up to 3 gallons with water; stir in 9 lbs. of dark gum substitute, and finally boil again.

2. LIGHT RED ALKALINE MORDANT.
    Alkaline red mordant, No. 1 . . 1 gallon.
    Dark gum substitute water . . 9 lbs.

3. ALKALINE PINK MORDANT.
    Potash lye at 54° Twaddle . . . 40 gallons.
    Sulphate of alumina (patent
      or cake-alum) . . . . . . 140 lbs.

Heat the lye in an iron boiler, and add the sulphate of alumina gradually, frequently stirring. The above proportions yield about 45 gallons at 34° Twaddle. It is to be thickened with dark gum substitute.

4. ALKALINE LIGHT PINK MORDANT.
    Potash lye at 41° Twaddle . . 25 gallons.
    Potash alum . . . . 90 lbs.
      Dissolve as in No. 3.

5. "Acid" (Lime-Juice Mixture).

    Lime-juice at 8° to 10° Twaddle   1 gallon.
    Starch . . . . 1 lb.
Boil, keeping it stirred till the starch is dissolved.

6. "Acid."

    Lime-juice at 20° Twaddle . . 1 gallon.
    Starch . . . . 1 lb.

7. "Acid."

    Lime-juice at 30° Twaddle . . 1 gallon.
    Starch . . . . 1 lb.

These acids are sometimes thickened with gum substitute.

8. Acid Discharge.

    Lime-juice at 22° Twaddle . . 1 gallon.
    Bisulphate of potash . . . 1 lb.
Decant the clear liquor, filter, and thicken with.
    Starch . . . . 1 lb.

9. Acid Discharge.

    Lime-juice at 28° Twaddle . . 1 gallon.
    Bisulphate of potash . . . 2 lbs.
Decant and filter the clear liquor, and thicken with
    Dark British gum . . . . 5 lbs.

10. Gall Liquor.

    Gall nuts, ground . . . . . 28 lbs.
    Acetic acid . . . . 2 gallons.
    Water 12   "

Let stand for two days, with occasional stirring, and then filter.

11. BUFF STANDARD.

  Water . . . . 2 gallons.
  Copperas . . . . 10 lbs.
  Brown acetate lead . . . . . 2½ ,,
  White ,, ,, ... 1¼ ,,

12. ANOTHER BUFF.

  Water . . . . 5 gallons.
  Copperas . . . . 20 lbs.
  Brown sugar lead . . . . . 10 ,,

Dissolve, and let settle, decant the clear liquid, and reduce it to the desired shade with gum Senegal water.

13. SULPHATE OF CHROME STANDARD.

  Water . . . . 6 gallons.
  Bichromate of potash . . . . 24 lbs.

Dissolve by means of heat, and place in a pan made of stoneware; then add

  Sulphuric acid, at 170° Twaddle 6½ pints.
  Cold water . . . . 3 gallons.

Then gradually add

  Sugar . . . . 6 lbs.

When the violent reaction and frothing cease, boil down to 3 gallons.

14. CHLORIDE OF CHROME STANDARD.

  Bichromate of potash . . . . 8 lbs.
  Boiling water . . . . 2 gallons.

Dissolve, and add

  Hydrochloric acid, at 32 Twaddle, 1 gallon 3¼ pints.

Then add gradually

  Sugar . . . . 3½ lbs.

15. **Arseniate of Chrome Standard.**

  Bichromate of potash . . . . . 10 lbs.
  Arsenious acid . . . . 14 „
  Hydrochloric acid, at 32° Twaddle 14 pints.
  Water . . , ... 2 gallons.

Heat the mixture until the liquid becomes of a pure green colour, without an olive tint. Sometimes, to obtain this result, more acid has to be used. When the reduction is completed, the liquid must be concentrated to 95° Twaddle. Care is necessary in the preparation of the chrome standards; should they not be perfectly uniform and smooth, they must be reheated, and more acid added if requisite. The best thickener for them is tragacanth gum.

Under Style 2, *Reserves*, will be found a variety of colours.

16. **Fast Blue Standard.**

  Ground-wet indigo . . . . 8 lbs.
  Soda lye at 70° Twaddle . . . $1\frac{1}{2}$ gallons.
  Water . . . . $1\frac{1}{2}$ „
  Feathered tin . . . . . . an excess.

Boil in an iron pan until a few drops placed on a piece of window glass appear perfectly yellow.

17. **Fast Blue for Block Work.**

  Fast blue Standard liquid (No.
   16) . . . . 1 quart.
  Tin crystals . . . . . . 12 ounces.
  Lime-juice, at 60° Fahr. (15·5° C.) 12 „
  Gum Senegal water (6 lbs. to the
   gallon) . . . . 3 quarts.

18. **Fast Green.**

  Fast blue Standard liquid (No.
   16) . . . . $2\frac{1}{2}$ pints.
  Lead gum (see No. 19) . . . 2 quarts.
  Tin crystals . . . . 8 ounces.

19. LEAD GUM.

    White sugar of lead . . . . 8 lbs.
    Nitrate of lead . . . . . . 4 ,,
    Hot water . . . . 1 gallon.
    Gum Senegal water (6 lbs. to
        the gallon) . . . . . . 1 ,,

20. DRAB.

    Sulphate of chrome, Standard
        No. 13 . . . . 5 quarts.
    Gum tragacanth water ($\frac{1}{2}$ lb. to
        the gallon) . . . . . . 10 ,,
    Cochineal liquor, at 4° Twaddle $1\frac{1}{2}$ pints.
    Bark liquor, at 8° Twaddle . . $1\frac{1}{2}$,,

21. FAWN.

    Sulphate of chrome, Standard
        No. 13 . . . . 1 gallon.
    Gum tragacanth water ($\frac{1}{2}$ lb. to
        the gallon) . . . . . . 2 ,,
    Brown standard, No. 5 . . . $\frac{1}{2}$ gallon.

22. PALE SAGE.

    Sulphate of chrome, Standard
        No. 13 . . . . 1 gallon.
    Gum tragacanth water ($\frac{1}{4}$ lb. to
        the gallon) . . . . . . 1 ,,

3. **Garancine Style.**—Most of the styles that can be carried out with madder can also be worked with garancine, but with this latter the resulting colours are generally not so clear. As a rule they are fuller, but less transparent, even when they have been equally well soaped and treated in the same manner after dyeing as the madders. The discovery and introduction of garancine was a great boon to

the calico-printer, since it not only enabled him to produce beautiful fast reds, purples, and other colours, cheaper than by the use of madder alone, but also chocolates, blacks, browns, drabs, &c., in conjunction with reds—results he was unable to get with madder. Furthermore, garancine and garanceux, can be used with peach-wood, yellow-wood, and Persian berry extract, and thus produce a fast print with red, yellow, brown, drab, chocolate, and black in combination; effects which can only be obtained with considerable difficulty and expense by the use of the madder-bath.

Garancines thus dyed are scarcely so fast as madder colours, nor will they tolerate soaping, but this operation is not so necessary as with madder dyed fabrics, for two reasons. In the first place the colours as they come out of the beck are generally brighter; and in the second, the whites or grounds are less tinged with the dye. The brightening of the colours and the clearing of the whites are effected by merely passing the dyed goods through what is called the "chlor" machine, or "chemicking" them. The "chlor" apparatus is a simple padding machine, immediately behind which is fixed a steam-box or iron chest, fitted with five copper rollers, and two perforated steam-pipes, also of copper. The trough of the padding machine contains solution of chloride of lime at from $\frac{1}{2}°$ to $1°$ Twaddle. The goods are passed through the trough and padding machine, and then immediately carried through a small aperture into the steam-box, and over and under the copper rollers; after which they pass through another aperture situated on the opposite side of the box, jets of steam being projected upon them during their passage. As they leave the steam-box, the pieces fall into a cistern of water, and then pass on to the squeezers. They are afterwards dried and finished.

From this it will be seen that the processes for dyeing and cleansing garancined goods are simpler and less expensive than those followed in madder dyeing. Garancine mordants also for reds, pinks, purples, and chocolates are used in a more dilute form than those for madder.

The same care, however, in drying the aluminous mordants after printing, is as necessary for garancine as for madder goods.

Formulæ for a dyed garancine print composed of black, red, brown, and drab, with chocolate ground.

BLACK. Same as No. 3, page 95.
RED. Same as No. 18, page 99.
BROWN. Same as No. 8, page 97.
CHOCOLATE. Same as No. 13, page 98.
DRAB. Same as No. 14, page 98.

4. **Padding Style.**—This style, as formerly practised, is now little used, being mostly confined to producing mourning effects, such as black and white, and black and lavender. The mordants employed in this style are the acetates of alumina and iron, the former giving with madder, garancine or alizarin reds; the latter, browns or blacks with logwood, and purples with madder and garancine.

By mixing the two mordants, and separately using for the tinctorial correspondents, logwood, quercitron bark, sumach and peach-wood, all the colours from clarets to olive can be obtained. The goods are dunged as for madders or garancines, and dyed and cleared mostly as garancines.

5. **Indigo Style.**—For Formulæ, see pages 89–90.

6. **China Blue Style.**—For Formulæ, see page 90.

7. **Indigo Discharge Style.**—The goods are dyed plain indigo, then soured, washed, dried, and afterwards printed. This very beautiful and permanent style, as more particularly exemplified in its chintz form, is of comparatively recent date. Instead of printing a reserve as in Style 5, and

afterwards dyeing and preparing, and finally blocking in the desired illuminating colours by block printing, the goods are first dyed (technically termed "dipped") plain, and afterwards printed with such compounds as discharge the indigo and leave the desired colours in the discharged parts. The application of chromic acid as the discharging agent used for this purpose, was first proposed by Mr. MERCER. The goods, after being dyed indigo and dried for printing, are previously padded through a solution, consisting of 8 to 12 oz. of bichromate of potash in a gallon of water, dried through a hot-air stove and excluded from light. They are then printed with a solution of oxalic and sulphuric acids, thickened with coloured starch (known to the calico-printer as "dark British gum"). After this, they are dried for printing in the usual way. They are then washed in water, drained, and again dried. There are other methods of conducting this process, but they are impracticable and more costly. We are indebted for the annexed formulæ, to Mr. JAMES CHADWICK of the firm of Messrs. CHADDERTON, CHADWICK & Co., of Manchester, who assures us their employment will be attended with complete success. The colours obtained by these formulæ, not only the white, but also yellow, orange, red, green, and, in short, almost all compound colours occurring as pigments, may be fixed. We may further state that this interesting style requires the observance of the two conditions of chemical and mechanical printing or fixing, to ensure the desired results, for whilst in the colours printed the chromic acid necessary for the discharge of the indigo must be present, they must also contain the coagulums or agglutinants necessary to fix the pigments, consequently it is necessary that the chromic acid and colour compounds should be as nearly as possible neutral.

After the goods are printed they are passed through the padding machine, before which is fixed a box containing five copper rollers, three at bottom and two at top. The fabric passes over and under these rollers, and then immediately after, through the padding machine, which causes the greater part of the solution with which the box and padding machine have been charged, to be squeezed out. The solution is composed of sulphuric acid at 12° Twaddle, containing in each gallon 8 oz. of oxalic acid. The temperature at which the solution must be maintained during the passage of the goods should be from 110° to 120° Fahr. (43° to 49° C.). After the goods or the piece passes from the padding machine, it should pass over one or two rollers so as to allow the gas to escape before the piece falls down in folds. At the expiration of a minute or two, the piece passes into the washing machine, and is lastly washed, dried and finished.

1a. WHITE DISCHARGE FOR INDIGOES.
   Water . . . . 2 gallons.
   Bichromate of potash . . . . 3 lbs.
   Dissolve,
   Dark British gum . . . . . 5 to 6 lbs.
       Boil and work hot.

2a. OXALATE AND CHROMATE STANDARD.
   Blood albumen solution (6 lbs.
       per gallon). . . . . . . 2 gallons.
   Neutral chromate of potash . . 8 lbs.
   Neutral oxalate of potash . . $1\frac{3}{4}$,,
       Dissolve perfectly cold; strain.

1. ORANGE FOR INDIGO DISCHARGE STYLE.
   Orange pigment . . . . . . 1 quart.
   Oxalate and chromate standard . 2 ,,
       Mix thoroughly and strain.

2. YELLOW DITTO.
   Yellow pigment . . . . . . 1 quart.

Oxalate and chromate standard . 1½ quarts.
Albumen thickening (6 lbs. per gallon) . . . . ½ ,,
  Mix thoroughly and strain.

3. GREEN DITTO.
Pigment green . . . . . . 4 quarts.
Oxalate and chromate standard . 4 ,,
  Mix thoroughly and strain.

4. RED DITTO.
Vermilion powder . . . . . 8 lbs.
Oxalate and chromate standard . 4 quarts.

Mix till thoroughly incorporated, and afterwards strain well.

5. BUFF (CHAMOIS).
Orange pigment . . . . . . 2 quarts.
Albumen thickening (6 lbs. per gallon) . . . . 12 ,,
Oxalate and chromate standard . 6 ,,
  Mix thoroughly and strain.

6. OLIVE DITTO.
Olive pigment . . . . . . 3 quarts.
Oxalate and chromate standard . 6 ,,
  Mix thoroughly.

7. BROWN DITTO.
Brown pigments . . . . . 4 quarts.
Oxalate and chromate standard . 8 ,,
  Mix thoroughly and strain.

8. LIGHT OLIVE DITTO.
Olive pigment . . . . . . 1½ quarts.
Oxalate and chromate standard . 4½ ,,
  Mix thoroughly and strain.

9. LIGHT BROWN DITTO.
   Pigment brown . . . . 2 quarts.
   Oxalate and chromate standard . 6 ,,
   Albumen thickening (6 lbs. per
     gallon) . . . . 2 ,,
       Mix thoroughly and strain.

10. SALMON, OR FLESH COLOUR, DITTO.
    Vermilion . . . . 4 lbs.
    Pigment brown . . . . 1 quart.
    Oxalate and chromate standard . 5 ,,
       Mix thoroughly and strain.

11. BLUE, DITTO (Light Blue.)
    Neutral Prussian blue paste . . 2 quarts.
    Oxalate and chromate standard . 4 ,,
       Mix thoroughly and strain.

N.B.—The brown and olive pigments are those usually sold for printing purposes. They are ground in water only, and must be perfectly neutral and insoluble. The ordinary green is a borate of chromium. Any pigment colour may be used that does not coagulate the albumen, which is mixed with it, and which will withstand the action of the sulphuric and oxalic acids when developing the design.

8. **Steam Style.** FOR COTTON, AND COTTON AND WOOL. (MIXED GOODS.) Preparation (before printing).—If the fabric consist of cotton only, the goods after being bleached and dried, are padded through a machine charged with stannate of sodium at 12° Twaddle, then immediately through sulphuric acid at 1½° to 2° Twaddle. This precipitates upon the fabric the tin oxide, which then becomes the mordant. The goods are next well and carefully washed, to free them from all traces of sulphuric acid; after which they are drained either by means of squeezers or the hydro-extractor. They are then dried for printing.

If the fabric is mixed wool and cotton (*mousseline de laine*), after the good pieces have been scoured or "crabbed," they are prepared as before described, but after being drained by means of the squeezer or the hydro-extractor, and before being dried, they are again padded through the padding-machine charged with sulpho-muriate of tin at 6° or 8° Twaddle, and allowed to remain so saturated from 1 to $1\frac{1}{2}$ hours. They are then passed through a cistern charged with chloride of lime at from $\frac{1}{2}$° to $\frac{3}{4}$° Twaddle. The cistern being supplied at top and bottom with rollers, the cloth passes over and under these; the cistern must have such a capacity that the cloth in its transit through it shall occupy from 30 to 50 seconds, during which time it is exposed to the action of the evolved chlorine. Directly the cloth leaves the chlorine cistern, it falls into water and thence passes on to the washing-machine, after which it is drained as usual, and dried.

Sulpho-muriate of tin is prepared as follows:—

    Sulphuric acid, 170° Twaddle . 2 gallons.
    Water . . . . 8 ,,

Stir till cool, then in 8 gallons of water dissolve, cold, protochloride of tin crystals . . . . 25 lbs.

Afterwards mix all together, and reduce with cold water to 8° Twaddle.

COLOURS FOR STEAM STYLES.

1. RED (wood red).

    Sapanwood liquor at 8° Twaddle $3\frac{1}{2}$ gallons.
    Ground alum . . . . . . 3 lbs.
    Chlorate of potash . . . . . 6 ozs.
    Nitrate of alumina . . . . . 3 pints.
    Bark liquor, at 8° Twaddle . . 5 ,,

CALICO-PRINTING.

    Water . . . . 8 pints.
    Crystal starch . . . . . . $7\frac{1}{2}$ lbs.
        Boil, cool, and strain.

2. NITRATE OF ALUMINA.
    Hot water . . . . 32 gallons.
In which dissolve
    Nitrate of lead . . . . . 96 lbs.
    Alum . . . . 96 ,,
    Common soda (carbonate) . . 12 ,,
When dissolved allow to subside and use the clear liquor.

3. PINK (cochineal) STANDARD.
    Boiling water . . . . . . 1 gallon.
In which dissolve
    Bitartrate of potash (cream of
      tartar) . . . . $1\frac{1}{2}$ lbs.
    Ground alum . . . . . . $1\frac{1}{2}$,,
And then add
    Cochineal liquor at 8° Twaddle 3 gallons.

4. PINK (cochineal) MEDIUM.
    Gum (foreign) water (4 lbs. to
      6 lbs. per gallon) . . . . 3 gallons.
    Cochineal pink standard (3) . 2 ,,

5. PINK (cochineal) PALE OR ROSE.
    Gum (foreign) water (5 lbs. per
      gallon) . . . . 3 gallons.
    Cochineal pink standard (3) . 1 ,,

6. PINK (magenta aniline) DARK.
    Gum (foreign) water . . . . 4 quarts.
    Magenta crystals . . . . . 4 ozs.
    Dissolved in
    Acetic acid at 8° Twaddle . . 1 quart.
    Tannic acid dissolved . . . . 12 ozs.

7. **Medium Pink.**

Foreign gum water	4 quarts.
Magenta crystals	2 oz.
Acetic acid at 8° Twaddle	1 quart.
Tannic acid dissolved	8 ozs.

8. **Pale or Light Pink.**

Foreign gum water	4 quarts.
Magenta crystals	1 oz.
Acetic acid	1 quart.
Tannic acid dissolved	6 ozs.

Steam Styles known as "best prepared Steams, not Anilines."

9. **Chocolate Steam Styles.**

Sapanwood liquor at 12° Twaddle	6 gallons.
Logwood " " "	3 "
Acetate of alumina 18° "	4 "
Bark liquor " 12° "	2 "
Starch	24 lbs.
British gum (torrified starch)	4 "
Alum	5½ "
Chlorate of potash	1¼ "
Sal ammoniac	2 "
Acetate of copper	1 gill (2⅛ noggins)

Boil, cool, and strain.

10. **Dark Blue (Royal).**

Water	6 gallons.
Starch	12 lbs.
Sal ammoniac	2 "

When boiled add

Prussiate of tin pulp (cyanide of tin)	6 gallons.

Thoroughly incorporate and add

Yellow prussiate of potash (ferro-cyanide of potassium)	12 lbs.

Red prussiate of potash (ferrid-
  cyanide of potassium) . . . 6 lbs.
Tartaric acid . . . . 18 „
Stir till quite dissolved; then add
Oxalic acid, which has previously
  been dissolved in 3 pints of
  water . . . . 1¼ lbs.
        Mix and strain.

11. MEDIUM BLUE.
    Gum substitute water . . . . 3 gallons.
    Dark royal blue . . . . . . 2 „

12. PALE BLUE.
    Gum substitute water . . . . 7 gallons.
    Dark royal blue . . . . . . 1 gallon.
Any further reduced shades in the same way to pale sky.

13. GREEN, (Dark Royal).
    Bark liquor at 12° Twaddle . . 6 gallons.
    Starch . . . . 10 lbs.
  Boil and add
    Protochloride of tin crystals . 12 ounces.
    Prussiate of tin pulp . . . . 1 gallon.
    Yellow prussiate of potash . . 14 lbs.
    Tartaric acid . . . . 13 „
    Extract of indigo (of commerce) 2 pints.

14. PRUSSIATE OF TIN PULP (Ferrocyanide of tin).
    Take—
    12 gallons of hot water, in which dissolve
    12 lbs. of yellow prussiate of potash; add gradually to
      the solution
    6 quarts protochloride of tin at 120° Twaddle.
  Then fill up the vessel with cold water; wash the preci-

pitate several times, and afterwards filter for use ; using of course the precipitate.

15. MEDIUM GREEN.
    Persian berry liquor (yellow berries) at 12° Twaddle . . 8 gallons.
    Yellow prussiate of potash . . 16 lbs.
    Alum . . . . 8 „
    Oxalic acid . . . . 2 „
    Protochloride of tin at 120° Tw. 2 „
    Acetic acid at 8° Twaddle . . 1 gallon.
    Foreign gum water (8 lbs. per gallon) . . . . 6 gallons.

16. PALE GREEN.
    Gum (foreign) water . . . . 5 gallons.
    Medium green . . . . . . 2 „

17. YELLOW.
    Persian berry liquor at 8° Tw. . 4 gallons.
    Starch . . . . 4 lbs.
    Gum substitute . . . . . . 1 „
    Alum . . . . 2 „
    Boil and add
    Protoxide of tin . . . . . . 4 pints.

Protoxide of tin is made as follows :—$2\frac{1}{2}$ lbs. protochloride of tin crystals are dissolved in 1 gallon of water, and $2\frac{1}{2}$ lbs. of common soda in another gallon, then both solutions are gradually mixed, and afterwards the precipitate washed twice and filtered.

18. DEEP YELLOW.
    Bark liquor at 12° Twaddle . . 4 gallons.
    Starch . . . . 6 lbs.
    Boil and add the following basic tin compound :—

Basic Tin Compound.

2½ lbs. of common soda are put into an earthenware vessel, which is then placed in a hot-water bath, and kept there till the heat from the water-bath causes the soda to lose its crystalline state, and to become liquid, when there is immediately added to it 3 lbs. of protochloride of tin crystals; the mixture must be briskly stirred, and when the whole has become semi-fluid, it is added to the colour.

19. RED (Wood Red) better than No. 1.
    Sapanwood liquor at 12° Twaddle   8   quarts.
    Bark liquor at 12° Twaddle . .   2½ ,,
    Nitrate of alumina   2   ,,
    Water   4   ,,
    Oxalic acid   1   lb.
    Alum . . . .   12 oz.
    Chlorate of potash   4   ,,
    Starch . . . .   6½ lbs.
            Boil, cool, and strain.

20. MEDIUM BROWN.
    Bark standard . . . . . .   14 pints
    Sapan   ,, ...   3   ,,
    Blue   ,, ...   1½ ,,
    Black   ,, ...   1   ,,
    Gum substitute water   3   ,,

21. BARK STANDARD (A).
    Bark liquor at 12° Twaddle . .   3 gallons.
    Alum . . . .   3 lbs.

22. SAPAN STANDARD (B).
    Sapanwood liquor at 12° Twaddle   2 gallons
    Chlorate of potash . . . . .   4 oz.

Alum . . . . 1½ lbs.
Gum substitute . . . . . . 8 lbs.
    Boil and cool.

23. BLUE STANDARD (C).

Water . . . . 2 gallons.
In which dissolve
 Red prussiate of potash . . . 2 lbs.
 Alum . . . . 1 ,,

24. BLACK STANDARD (D).

 Logwood liquor at 12° Twaddle . 2 gallons.
 Gum substitute . . . . . . 6 lbs.
Boil and add
 Red prussiate of potash . . . 2 lbs.
 Alum . . . . 1½ lbs.
    Dissolve and cool.

25. RED BROWN.

 Bark standard (A) . . . . . 8 quarts.
 Sapan ,, (B) . . . . . 2½ ,,
 Black ,, (D) . . . . . 1 ,,
 Gum substitute water . . . . 6 ,,

26. FAWN or MEDIUM (CHAMOIS).

 Bark standard (A) . . . . . 16 pints.
 Sapan ,, (B) . . . . . 2 ,,
 Black ,, (D) . . . . . 2 ,,
 Gum substitute water . . . . 16 gallons.

N.B.—From the four standards A, B, C, and D, every tone and shade of colour may be obtained, from a dark brown to a light fawn (including Chestnut Cuir, Chamois, Cinnamon and all kindred compound colours), by varying their proportions.

27. BUFF or PALE NANKEEN STANDARD (A).

 Persian berry liquor at 12° Twaddle 3 gallons.

CALICO-PRINTING.

  Cochineal liquor at 6° Twaddle 1 gallon.
  Alum . . . . 3 lbs.
  Bitartrate of potash (cream of
   tartar). . . . . $1\frac{1}{2}$,,

28. MEDIUM BUFF.
  Gum substitute water . . . 3 gallons.
  Buff standard (A). . . . . . 2 ,,

29. PALE BUFF.
  Gum substitute water . . . 10 gallons.
  Buff standard (A) . . . . . 1 ,,

30. BLUE STANDARD FOR SHADES.
  Water . . . . 4 gallons.
  In which dissolve
  Oxalic acid . . . . 1 lb.
  Alum . . . . 4 ,,
  Yellow prussiate of potash . . 8 ,,

31. SLATE STANDARD.
  Logwood liquor at 12° Twaddle 2 gallons.
  Blue Standard for shades . . $1\frac{1}{4}$,,
  Medium green . . . . . 3 quarts.
  Tartaric acid . . . . . . 1 lb.

32. MEDIUM SLATE.
  Gum substitute water . . . 4 quarts.
  Slate standard, No. 31. . . . 1 ,,

33. PALE SLATE.
  Gum substitute water . . . . 10 gallons.
  Slate standard . . . . . . 1 ,,

34. SILVER DRAB STANDARD.
  Slate standard . . . . . . 4 gallons.
  Buff or pale nankeen standard . 3 ,,
  Pink (cochineal) standard (3) . $\frac{1}{2}$,,

35. **Medium Silver Drab.**
    Gum substitute water . . . . 3 gallons.
    Silver drab standard . . . . 1 „

36. **Pale Silver Drab.**
    Gum substitute water . . . 10 gallons.
    Silver drab standard . . . . 1 „

N.B.—From the foregoing formulæ every shade and tone of compound colours can be obtained:—Stones, Drabs, Lavenders, Slates, Pearls, &c. &c., all of which are known as "best prepared steam styles."

FORMULÆ FOR MIXED GOODS.

Cotton and Woollen (Mousseline de Laine), &c.

On prepared goods.

1. **Red.**
    Cochineal liquor, at 12° Twaddle   4 gallons.
    Starch . . . . 8 lbs.
    Gum substitute . . . . . . 1 „
    Boil and add
    Protochloride of tin crystals . . 1½ „
    Oxalic acid . . . . 1½ „
        Dissolve and strain.

2. **Crimson.**
    Ammoniacal cochineal standard . 4 gallons.
    Bitartrate of potash (cream of
        tartar) . . . . 3 lbs.
    Alum (dissolved) . . . . . 3 „
    Foreign gum . . . . 16 „
        Stir till dissolved.

3. **Rose or Pink.**
    Foreign gum water . . . . 3 gallons.
    Crimson, No. 2 . . . . 1 „

4. **Ammoniacal Cochineal Standard.**

    8 lbs. Cochineal, steeped for 12 hours in a covered vessel (not copper) with
2 gallons cold water, and
1 ,,   liquid ammonia; afterwards add
2 ,,   more water; boil in water-bath for 2 hours, afterwards sieve off the liquor, which ought to measure 4 gallons; should it not do so, add sufficient water to make up to that quantity.

5. **Dark Blue.**

    Same as royal blue for cottons.

6. **Light Blue.**

    Same as pale blue for cottons.

7. **Yellow or Amber.**

Persian berry liquor, at 12° Twaddle	4 gallons.
Foreign gum	16 lbs.
Protochloride of tin crystals	2 ,,

8. **Green Medium.**

Persian berry liquor, at 12° Twaddle	4 gallons.
Alum	3 lbs.
Oxalic acid	1 ,,
Tin crystals (protochloride)	1 ,,
Yellow prussiate of potash	6 ,,
Foreign gum water (6 lbs. per gallon)	4 gallons.
Extract of indigo	3 pints.

9. **Chocolate.**

Sapanwood liquor, at 12° Twaddle	6 gallons.
Logwood liquor, at 12° Twaddle	3 ,,

Bark liquor, at 12° Twaddle . . 1 gallon.
Starch . . . . 16 lbs.
Gum substitute . . . . . . 4 ,,
Boil and add
Alum . . . . 7½ ,,
Chlorate of potash . . . . . 10 ozs.
Red prussiate of potash . . . 4½ lbs.

Most of these colours are now superseded by the aniline colours.

9. **Spirit Styles.**

1. BLACK.
Logwood liquor at 8° Twaddle . 6 gallons.
Starch . . . . 9 lbs.
Gum substitute . . . . . . 4 ,,
Boil and add
Nitrate of iron at 84° Twaddle . 4 pints.
Mix thoroughly and strain.

2. PURPLE or LILAC.
Logwood liquor at 8° Twaddle . 8 gallons.
Starch . . . . 16 lbs.
Boil and add
Yellow prussiate of potash . . 1¼ ,,
Nitrate of iron at 84° Twaddle . 3½ pints.
Starch paste (1 lb. per gallon) . 4 gallons.
Perchloride of tin 120° Twaddle 1 ,,
Mix and Strain.

3. PINK. Use Sapanwood, with tin.
4. BLUE. Use Prussian blue, with tin.
And so on.

10. **Bronze Style.***—The goods are padded through a

---

* During the last three years there has been more demand for prints produced by this style, than for the previous forty years.

padding machine, charged with a solution of protochloride of manganese at from 24° to 36° Twaddle, and dried through a hot-air stove, care being taken to avoid their coming into contact with the iron of the stove. They are then passed through the soda cistern, as quickly as possible after being dried, since the protochloride of manganese being deliquescent, the fabric otherwise becomes damp, to the detriment of the bronze colour. The soda cistern is fitted with twelve rollers, six at top and six at bottom, over and under which the cloth is passed. The cistern is charged with caustic soda at from 20° to 24° Twaddle. In passing through this solution, the manganese is precipitated upon the fabric, in an almost colourless condition, in the state of a protoxide.

The fabric is then carried over wooden rollers, a process which, by exposing it to the air, effects the partial oxidation of the manganese. It is then well washed and afterwards winched for five or ten minutes in chloride of lime at 4° to 6° Twaddle, which converts the manganese protoxide into peroxide, which is of a dark bronze colour, and which gives its name to the style. After the pieces are printed with the colours, for which the formulæ are now given, they are steamed in the usual way in an ordinary steam-chest at about one to two pressures for twenty-five minutes. They are afterwards washed, drained on a hydro-extractor, and dried.

COLOURS FOR BRONZE STYLES.

1. WHITE DISCHARGE ON BRONZE.

  Water . . . . 2 gallons.
  Wheaten starch . . . . . . 5 lbs.
 Boil and, when half cold, add
  Protochloride of tin crystals . . 12 ,,

Stir till thoroughly dissolved, and afterwards strain.

2. **Yellow Ditto.**
   Bark liquor, at 12° Twaddle . . 2 gallons.
   Starch . . . . 5 lbs.
   Boil and add when half cold
   Protochloride of tin crystals . . 12 ,,
   Dissolve and strain.

3. **Red Ditto.**
   Sapan or Brazil wood liquor, at
     12° Twaddle . . . . . . 2 gallons.
   Starch . . . . 5 lbs.
   Boil and, when half cold, add
   Protochloride of tin crystals . . 12 ,,
   Dissolve and strain.

4. **Mauve or Violet Discharge on Bronze.**
   4 B Violet . . . . 2 quarts.
   Water . . . . 4 gallons.
   Starch . . . . 11 lbs.
   Boil and, when half cold, add
   Protochloride of tin crystals . . 20 ,,
   Dissolve and strain.

5. **Blue Ditto.**
   Prussian blue paste . . . . 2 quarts.
   Water . . . . 2 gallons.
   Starch . . . . 6 lbs.
   Boil and, when half cold, add
   Protochloride of tin crystals . . 12 ,,
   Dissolve and strain.

6. **Green Ditto.**
Mix yellow No. 2 and blue No. 5, say, 2 of the former, $1\frac{1}{2}$ of the latter, or any other proportions to produce the shade of green required.

From the foregoing formulæ, it will be seen that protochloride of tin is the active agent which discharges the peroxide of manganese, and at the same time becomes the mordant for the various colouring matters. Any colouring matter therefore to which the protochloride of tin acts as a mordant, and which will bear the quantities given of it as before, can be used for bronze discharges.

Beautiful styles are now being worked with bronze effects.

11. **Pigment Style.**—The theory and practice are described at page 93. Sometimes, however, to obtain bright blues, the goods instead of being steamed are passed through a cistern charged with boiling milk of lime. The hot lime-water, by instantly coagulating the egg albumen, renders the blue brighter. Sometimes after fixation of the pigments, they are slightly soaped with the object of removing the unpleasant odour.

PIGMENT COLOURS.

1. GREEN (Best).
     Green pigment . . . . 8 lbs.
     Blood albumen water (5 lbs. per
        gallon) . . . . 1 gallon.
         Mix well and strain.

2. GREEN (Generally Used).
     Green pigment . . . . 4 lbs.
     Gum tragacanth water . . . 1 quart.
     Blood albumen water (5 lbs. per
        gallon) . . . . 3 ,,

3. BROWN PIGMENT COLOUR.
     Brown pigment . . . . 4 lbs.
     Gum tragacanth water. . . . 1 gallon.
     Blood albumen water . . . . 1 ,,

4. BUFF PIGMENT COLOUR.
    Buff pigment . . . . 4 lbs.
    Gum tragacanth water . . . 2 quarts.
    Blood albumen water . . . . 2 ,,

5. BLACK PIGMENT COLOUR.
    Pigment black . . . . . . 6 lbs.
    Gum tragacanth water . . . 2 quarts.
    Blood albumen . . . . . . 4 ,,

6. OLIVE PIGMENT COLOUR.
    Pigment green colour . . . . 1½ pints.
    Buff pigment   ,, ... 1½ ,,
    Black   ,,   ,, ... 2 ,,
    Blood albumen water . . . . 2 ,,
    Gum tragacanth water . . . 2 ,,

7. TAN PIGMENT COLOUR.
    Blue pigment standard . . . 1½ pints.
    Brown   ,,   colour . . . 1½ ,,
    Black   ,,   ,, ... 1½ ,,
    Blood albumen water . . . . 2 ,,
    Gum tragacanth water . . . 2 ,,

8. GREY PIGMENT COLOUR.
    Black pigment colour . . . . 2 pints.
    Albumen water . . . . . . 1 ,,
    Gum tragacanth water . . . . 1 ,,

9. SLATE PIGMENT COLOUR.
    Black pigment colour . . . . 3 pints.
    Blue   ,,   standard . . . 12 ,,
    Blood albumen water . . . . 7 ,,
    Gum tragacanth water . . . . 7 ,,

10. PIGMENT BLUE STANDARD.

Ultramarine . . . . 2 lbs.
Gum tragacanth water . . . 3 quarts.
Blood albumen water . . . . 1 ,,

The best pigment blues are made with egg albumen, not with blood albumen.

12. **Extract Style.**—Some years ago, before the adoption of artificial alizarin by the dyer and calico-printer, an extract of madder was introduced into the trade. It is from this fact that the present style takes its name. By the extract style, thanks to the discovery of artificial alizarin and other coal-tar colours, effects are obtained that were previously practically impossible. The old method was to print on the fabric the mordants for the madder colours, and afterwards to dye it, the printing in of the illuminating colours being performed by hand. In extract printing, this clumsy method, which generally effaced the integrity of the design, is avoided, with immense advantage to the resulting pattern, of delicacy and fidelity of outline, as well as of richness and purity of colour. The discovery and application of the coal-tar colours have given an unprecedented impetus to this amongst other branches of calico-printing, not only by increasing the number and variety of tinctorial agents possessing purer tints, but because the raw material for manufacturing them lies around us. At the present day when the printer requires pale chintzes, it is not necessary for him to ransack the two hemispheres for the red, yellow, blue, green, and violet dye-stuffs, as it was not many years back.

COTTON.

FORMULÆ FOR EXTRACT STYLES.

After the goods are bleached as for madder styles, they are prepared by padding through an emulsion of oleine, or saponified castor oil, made by mixing 1 part of oleine to 15 of water; or 1 of oleine to 20 of water: this latter being the strength generally used. The goods are afterwards dried, and are then ready for printing.

After printing, the goods are aged and then steamed for 1½ hours at a pressure of from 1 to 2 lbs. per square inch. Then, provided the prints contain no mordant of which tannic acid is an ingredient, as is the case with the coal-tar colours, they are washed and afterwards soaped as in madder work, but for not quite so long a time, or at so high a temperature; afterwards they are washed, cleared, drained, and dried. If, however, the mordants used for any of the colours contain tannic acid, the goods must be passed through a solution of tartrate of antimony (2 oz. to the gallon). This is done before washing and soaping, and of course after they have been steamed.

1. EXTRACT, RED.
    Acetic acid at 8° Twaddle . . 9 pints.
    Water . . . . 14 ,,
    Olive oil . . . . 4 ,,
    Starch . . . . 7 lbs.
Boil, and when nearly cold, add
    Acetic acid at 8° Twaddle . . 2 pints.
    Acetate of alumina at 18° Tw. . 2½ ,,
    Nitrate of alumina . . . . . 1½ ,,
    Sulphocyanide of alumina . . . 2 ,,
    Acetate of lime (at 2 lbs. per gal.) 5 ,,
    Artificial alizarin (at 20 per cent.) . . . . 13 lbs.

2. DARK PURPLE (Extract Work).
    Water . . . . 2 gallons.
    Acetic acid at 8° Twaddle . . 2½ quarts.
    Starch . . . . 6 lbs.

Boil and add
- Acetate of iron (4 and 4)* . . 2 pints.
- Acetate of lime (2 lbs. per gal.) 4 „
- Artificial alizarin (20 per cent.) 10 „

3. BLACK (Chromium).
- Logwood liquor at 12° Twaddle . 6½ gallons.
- Bark   „   „   „   . 6 quarts.
- Acetic acid at 8° Twaddle. . . 5 „
- Water . . . . 2 „
- Dark British gum . . . . 36 lbs.
- Starch . . . . 9 „

Boil and add
- Chlorate of potash . . . . . 1 „

And when cold add
- Nitro-acetate of chrome . . . 1 gallon.

4. NITRO-ACETATE OF CHROME.

A.

6 lbs. bichromate of potash,
3 gallons of hot water; dissolve, then add
7½ lbs. sulphuric acid at 170° Twaddle, diluted with
2 quarts of cold water; then add gradually
1¼ lbs. raw sugar.

B.

9¾ lbs. nitrate of lead,
9¾ „ acetate of lead dissolved in
2 quarts of hot water; then add to solution A. Mix well, allow to subside, and use the clear liquor.

---

* The acetate of iron is made as follows:—Sulphate of iron (copperas) 4 lbs., hot water 2 quarts, dissolve; acetate of lead 4 lbs., hot water 2 quarts, dissolve. Mix the solutions and stir well; let the precipitate settle, and keep the clear liquid for use.

5. MAUVE (Fast) for Extract Work.
    Acetic acid . . . . . . 2 quarts.
    Water . . . . 2 ,,
    Violet crystals (6 B violet.) . 4 ozs.
    Starch . . . . 1 lb.
Boil and add
    Tannic acid . . . . . . 8 ozs.
        Dissolve and strain.

6. MAUVE (Fast).
    Red liquor at 18° Twaddle . . 3 gallons.
    Water . . . . 1 ,,
    Gum tragacanth water (8 ozs.
      per gallon) . . . . . . 1 gallon.
    Starch . . . . 5 lbs.
Boil and add
    Alum . . . . 1 lb.
    Violet crystals . . . . . . 1 ,,
Dissolve and, when cold, add
    Glycerin standard . . . . . $1\frac{1}{2}$ quarts.
        Mix well and strain.

7. GLYCERIN STANDARD.
    Brown glycerin . . . . . . 2 gallons.
    Arsenic . . . . 8 lbs.

Boil 1 hour; allow to stand, and use the clear liquor.

P.S.—When boiling the arsenic and glycerin, avoid inhaling the steam, which contains a dangerous gas.

N.B.—By following Methods 5 and 6 all the aniline violets, mauves, magentas, most of the blues, chocolates, and other aniline colours can be fixed for printing; when desired for dark shades and neat patterns, starch may be used as the thickening, and gums for medium or light shades.

8. CERULEAN BLUE (most beautiful colour.)
    Red liquor at 18° Twaddle . . 3 gallons.
    Water . . . . 1 „
    Gum tragacanth water (8 ozs. per
        gallon) . . . . 1 „
    Starch . . . . 5 lbs.
  Boil and add
    Ground alum . . . . 1 lb.
  Dissolve and, when quite cold, add,
    Glycerin standard . . . . . 3 pints.
  And
    Cerulean blue (ROBERTS, DALE,
        and Co.) . . . . 5 lbs.

9. METHYL GREEN.
    Acetic acid . . . . 7½ quarts.
    Sumach extract . . . . . . 7½ „
    Tartaric acid . . . . 1 lb.
    Starch . . . . 5 „
  Boil and, when cold, add
    Methyl green crystals . . . 1 „
    Alum . . . . 8 ounces.
  Dissolved in
    Persian berry liquor or extract,
        at 48° Twaddle . . . . . 4 pints.

10. METHYLINE BLUE.
    Acetic acid at 8° Twaddle . . ½ gallon.
    Water . . . . ½ „
    Methyline blue crystals . . . 3 ounces.
    Starch . . . . 1 lb.
  Boil and add
    Tannic acid . . . . 8 ounces.
            Dissolve.

11. PINK or ROSE.

Is made by reducing Extract Red, formulæ (1), either with gum water or starch paste.

12. PINK.
    Gum water (4 lbs. per gallon) . 3 gallons.
    Extract Red (1) . . . . . 1 ,,

13. PALE PURPLE.
    Gum water (4 lbs. per gallon) . 3 gallons.
    Extract Purple (2) . . . . 1 ,,

CHAPTER IV.

DYE STUFFS.

Aloes.—An extract of aloes when treated with nitric acid, gives rise to various beautiful coloured products, which, by the aid of mordants, can be fixed to silken and woollen goods. *Aloin*, the colour-giving principle of aloes, is a body soluble both in water and alcohol, and when exposed to the air, it absorbs atmospheric oxygen, and assumes an intense red colour. Aloes are seldom employed for dyeing purposes.

Annotta. *Syn.* ANOTTO, ANNATTO, ANNATA, ARNATTO, ARNOTTO.—A colouring matter obtained from the seeds of the *Bixa orellana* (LINN.), an exogenous evergreen tree, common in Cayenne, and some other parts of tropical America. Annotta is usually obtained by macerating the crushed seeds or seed-pods of the plant in water for several weeks, ultimately allowing the pulp to subside, then boiling it in coppers to a stiff paste, and drying it in the shade. Sometimes a little oil is added when the paste is made into cakes or lumps. A better method is that proposed by LEBLOND, in which the crushed seeds are simply exhausted by washing them in water (alkalized?), from which the colouring matter is afterwards precipitated by means of vinegar or lemon-juice; the precipitate being subsequently collected, and either boiled up in the ordinary manner, or drained in bags and dried, as is in the preparation of indigo. Annotta so prepared is said to be four times as valuable as that made by the former process.

The term annotta is frequently indiscriminately applied to the commercial article, and to the colouring principle contained in it. This latter is a resinous substance possessing strong tinctorial properties, to which the name *bixin*[*] has been given. It may be prepared by digesting commercial annotta in an alkaline lye, and by neutralizing the filtered liquid with sulphuric acid, when the bixin is precipitated. Bixin is of an orange colour, scarcely soluble in water, but freely so in alcohol, ether, oils, and fats, to each of which substances it imparts a beautiful orange tint. It is also soluble in alkaline solutions, to which it imparts a deep red colour.

Genuine commercial annotta contains about 28 per cent of the resinous substance (bixin), and 20 per cent. of extractive matter.

In analyzing a sample, it is only necessary to determine the quantity of ash and of colouring matter it contains, the nearer of course the amount of the latter approaches that given above, the greater will be its trade value.

Dr. BLYTH gives the following as the composition of a fair commercial sample. The sample was in the form of a paste, colour deep red, odour peculiar, but not disagreeable :—

Water	24·2
Resinous colouring matter	28·8
Ash	22·5
Starch and extractive matter	24·5
	100·0

The following is an analysis of an adulterated specimen. The sample was in a hard cake of a brown colour, with the

---

[*] Annotto also contains another and less important colouring body, which has been denominated *orellin*.

maker's name stamped upon it, and marked "patent"; texture hard and leathery, odour disagreeable :—

Water	13·4
Resin	11·0
Ash, consisting of iron, chalk, salt, alumina, silica	48·3
Extractive matter	27·3
	100·0

Thus, in the one the resin was 28 per cent., the ash 22 per cent.; in the other the resin was only 11 per cent., the ash no less than 48 per cent.

Annotta is very frequently extensively adulterated. The most usual sophisticants are meal, flour, or farina of some description, chalk, plaster of Paris, pearlash, soap, turmeric, Venetian red, red ochre, orange chrome and common salt. Red lead and sulphate of copper have occasionally been detected in it. Dr. HASSALL states that out of thirty-four different specimens, two only were genuine. Since genuine commercial annotta exhibits but few evidences of structure, any of the above vegetable adulterants may be easily detected by means of the microscope.

Annotta is used as a pigment for painting velvets and transparencies, and as a dye-stuff for cotton, wool and silk, to the latter of which it imparts a beautiful orange yellow hue, the shade of which may be varied from "aurora" to deep orange, by using different proportions of pearlash with the water in which it is dissolved; or by applying different mordants before adding the annotta to the dye-beck. The tints thus imparted are, however, more or less fugitive.

Archil. *Syn.* ARCHEL, ORCHIL.—This is a violet-red, purple or blue colouring matter or dye-stuff, obtained from several species of lichens, but of finest quality from *Rocella tinctoria*, and next from *Rocella fusiformis*.

The archil of commerce is met with as a liquid paste, or as a thin liquid dye of more or less intensity of shade. Blue archil, is prepared by steeping in the cold in closely covered wooden vessels, the coarsely ground lichen in a mixture of lime or milk of lime, in stale urine or bone spirit, or in any similar ammoniacal solution; the process being repeated till all the colour is extracted. The carbonate of ammonia resulting from the decaying urine acts upon the peculiar acids, the lecanoric, alpha and beta orcellic, erythrinic, gyrophoric, evernic, usninic, &c., contained in the lichens, and converts them into *orcine*. By taking up nitrogen and oxygen, *orcine* is converted into *orceine*, which constitutes the essential colouring principle of archil. In the preparation of red or crimson archil, the same materials are used as for the production of the blue; the only difference being that a smaller quantity of milk of lime is used, and the steeping is generally performed in an earthen jar placed in a room heated by steam, and technically called a stove. The two varieties differ only in the degree of their red or violet tint, the addition of a small quantity of lime or alkali to the one, or of acid to the other, immediately bringing them to the same shade of colour. The hues given by archil to silk and wool possess an exquisite lustre, but they are far from permanent, and since the introduction of the coal-tar colours, their use has diminished considerably. If archil is employed at all, it is in combination with other dye stuffs, or as a finishing bath to give a bloom to silk or woollen goods dyed with some permanent colour.

Barwood.—A red dye-wood imported from Angola and other parts of Africa. It closely resembles camwood and sanders wood in its colouring matter being of a resinous nature, and scarcely soluble in water. In dyeing, this difficulty is obviated by taking advantage of the strong affinity existing between it and the protosalts of tin and iron. Thus, by strongly impregnating the goods with protochloride of tin, either with or without the addition of sumach, according to the shade of red desired, and then putting them into a boiling bath containing the rasped wood, the colour is rapidly given out and taken up, until the whole of the tin in the fibres of the cloth is saturated, and the goods become of a rich bright hue. In like manner the dark red of bandana handkerchiefs is commonly given by a mordant of acetate of iron followed by a boiling bath of this dye-stuff. Previous to the introduction of artificial alizarin "Barwood Reds" were extensively used, and they ranked next in permanency to madder reds. The dark barwood red, however, possessed one great defect : if exposed for some time to the air, it darkened, and became dull. The change is supposed to be due to the absorption of ammonia.

Catechu. *Syn.* CASHEW, CUTCH, GAMBIR. — An extract obtained from the wood of the *Acacia catechu*, or from the leaf of *Uncaria gambir*. There are several varieties of catechu known in commerce, of which the principal are :—

BOMBAY CATECHU.—This occurs as a firm, brittle extract, of a dark-brown colour, of uniform texture, and of a glossy, semi-resinous and uneven fracture. Sp. gr. 1·39. Richness in catechu tannin 52 per cent.

MALABAR CATECHU.—Resembles the last in appearance, but is more brittle and gritty. Richness in catechu tannin 45·5 per cent.

The amount of catechu tannic acid may be determined in catechu as follows :—

1. Exhaust a weighed and finely pulverized sample of the catechu with ether, and evaporate by the heat of a water-bath; the product, which is catechu-tannin, must then be accurately weighed.

2. Reduce the sample to powder, dissolve in hot water, let cool out of contact with the air, filter and add solution of gelatine as long as a precipitate falls. The precipitate, after being washed and dried at a steam heat, should contain 40 per cent. of catechu-tannin.

When used for dyeing purposes catechu forms a great variety of browns.

Alum mordants are mostly employed in dyeing with it. With the salts of copper, and sal ammoniac,* catechu gives a fast bronze colour; with protochloride of tin, a brownish yellow; with perchloride of tin and nitrate of copper, a deep bronze; with acetate of alumina, a reddish brown; with nitrate of copper a reddish olive grey, and with nitrate of iron, a deep brown grey.

Acetate of alumina and tin nitrates make but weak mordants for catechu, sulphate and acetate of copper the best. The iron mordants also act satisfactorily, although they give darker and duller colours than the copper ones. Hence it is that the iron mordants are rarely ever used alone with catechu, but in conjunction with copper ones, by which means the browns are converted into drabs. Some beautiful colours are now obtained by mixing varying quantities of magenta with the catechu and copper. Such colours are fairly fast.

Brazil Wood. *Syn.* CAMWOOD. — This dye-stuff is furnished by several species of trees belonging to the genus *Cæsalpinia*, and was formerly much employed in producing various shades of red. The best kind of Brazil wood is that known as Pernambuco or Fernambuco wood, the source of which is the *Cæsalpinia brasiliensis s. crista*.

This variety is a heavy and rather hard wood, externally

---

* The sal ammoniac acts chiefly by absorbing moisture, and by thus expediting the oxidation of the metals employed, aids in the fixation of the catechu.

DYE STUFFS.

of a yellowish brown, and internally of a bright red colour. It occurs in commerce in chips and large logs. Inferior but closely allied varieties of the wood are—1. *Sapan wood*, derived from *Cæsalpinia Sapan*, and *Lima* or *Nicaragua wood*.

The colouring matter of all these woods is *brezilin*, which when isolated occurs in small orange-coloured needles, soluble both in water and alcohol. Alkalies turn it violet, acids yellow. These woods give brilliant but unstable colours.

Coal-tar Colours.—See p. 160.

Cinchonine.—If cinchonine (one of the products left after the extraction of the quinine from cinchona bark),. be submitted to distillation with caustic soda in excess, there passes over into the receiver a crude kind of oil, called chinoline oil, one of the principal constituents of which is a base called *Lepidine*.

When this chinoline oil is heated with amyl iodide, and the product treated with caustic soda solution, a very brilliant blue pigment known as *Cyanine, Lepidine Blue,* or *Chinoline Blue* is obtained.

Cinchonine has a very limited application in dyeing.

Cochineal. *Syn.* COCCUS CACTI.—This insect is found upon several species of *Cacti*, more particularly on the Nopal plant, and on the *Cactus opuntia*. The chief seats of the cochineal culture are Mexico, Central America, Java, Algeria, the Canary Islands and the Cape. The insect sickens and dies out if the cactus plant is grown too near the sea, or exposed to damp winds.

The female insect, which only possesses value as a dye, is wingless; the male, on the contrary, is winged. The females are collected twice a year after they have laid their eggs. They are first brushed off the plant, and then killed, sometimes by exposure to the vapour of boiling water, but more frequently by the heat of an oven.

The two chief varieties of cochineal are known in commerce as the "silver grey" and the "black," another kind is dark grey mottled with red. The best kind comes from Honduras. An inferior quality is collected from wild cactus plants. The silver grey specimens are covered with a white dust, which microscopical examination has proved to be the insect's excrement. The white dust is frequently imitated by shaking the insects in a bag with French chalk, or white lead. Herr DURWELL, a German chemist, states that he found a sample adulterated with oxide of zinc. Sulphate of baryta and powdered bone dust have also been used as sophisticants. The object of the adulterations is of course to increase the weight. Genuine cochineal has the specific gravity 1·25. A peculiar kind of acid, which has been named carminic acid, has been discovered in cochineal. When acted upon by very dilute sulphuric acid and other reagents, carminic acid splits up into carmine red (or carmine) and glucose. Artifically prepared carmine is obtained by exhausting cochineal with boiling water, adding alum to the clear supernatant liquid, and allowing the carmine to deposit. Since the introduction of the coal-tar colours, cochineal is in much less demand by the dyer.

To ascertain the colorific value of a sample of cochineal, the dyer generally makes a dyeing experiment on a small scale. He impregnates a piece of mordanted wool, cotton, or silk, with a decoction of the specimen under examination, and then compares the colour with that of a piece of similar tissue dyed with a standard decoction of cochineal. By this means, he is enabled to form an opinion as to the strength, purity of colour, &c., of his specimen.

Cudbear. *Syn.* PERSIO.—This dye-stuff is obtained from *Lecanora tartarea*, and other lichens, by a process nearly similar to that used in making archil. The lichen is watered with stale urine or some other ammoniacal liquid, and kept in a state of fermentation for three or four weeks, after which the mixture is transferred to a flat vessel, and exposed to the air until the urinous smell has disappeared, and it has become of a violet colour. The residue is then ground to powder. Its use is limited to a few cases of silk dyeing, where it is employed to impart shades of ruby and maroon. It dyes wool of a deep red tint. The colours given by it are very fugitive; there are two varieties of cudbear, the blue and the red.

Fustic.—Two distinct dye-stuffs are met with under this name, the "old" and the "young" fustic.

Old Fustic, called also "yellow wood" is the hard wood of the *Maclura tinctoria*, a tree growing in Cuba, Hayti, and St. Domingo. The tinctorial properties of the wood are due to a colourless crystalline substance, called *morine*, and to a peculiar acid, the *moritannic*, to which the name *maclurin* has been given. Under the combined influence of alkalies and the air, morine becomes yellow. It dyes woollens different shades of yellow according to the mordant. These colours are very permanent. A commercial extract of the wood is sent into the market under the name of Cuba extract.

Young Fustic is the wood of the *Rhus cotinus* or *Venice sumach*, a shrub belonging to Southern Europe. It derives its name of young fustic from the circumstance of its branches being much smaller than those of the old fustic. The colouring principles of this wood are tannic acid, and a substance termed *fustine*, which, it appears, yields by decomposition *quercetin*, one of the decomposition products of *quercitrin*, the pigment of quercitron bark. Young fustic dyes greenish yellow, but the colours are not very permanent.

Indigo.—This blue dye-stuff is extracted from several plants growing in the East and West Indies, Central and Southern America, Egypt and other countries.

The *Indigofera tinctoria, Indigofera anil, Indigofera disperma, Indigofera pseudotinctoria,* and *Indigofera argentea* are the chief varieties of the plant, from which commercial indigo is obtained. The *Nerium tinctorium* is the source of the East India indigo. Indigo does not exist as such in the plant, but is the result of the action of atmospheric oxygen upon the freshly expressed juice.

The method of its manufacture consists in steeping the branches, twigs and leaves of the plant in tanks filled with water until fermentation sets in. The clear liquid, which then assumes a yellow or golden colour, is drawn off from the deposited vegetable matter, and agitated and beaten with bamboo poles for about two hours to bring it into contact with the air. By this treatment the indigo forms and settles down as a blue precipitate, which in its fluid state is run into a cauldron, in which it is boiled for about fifteen or twenty minutes to prevent its undergoing a second fermentation, which would render it useless. After standing over night, the magma in the cauldron is again boiled for three or four hours, after which it is placed on filters, composed of bamboo mats and canoes. The thick nearly black paste which is left on the filters, is subjected to pressure in boxes, whereby the greater portion of the water is removed, and the paste becomes of a more solid consistence. The cakes of indigo thus formed are dried by artificial heat, packed in wooden boxes and so sent into the market. Commercial indigo contains indigo blue, or *indigotin* (its most important constituent), indigo red, indigo brown, &c. The amount of indigo blue or indigotin varies in different samples of commercial indigo from 20 to 75 or 80 per cent., and averages from 40 to 50 per cent.

The indigo plant, according to SCHUNCK, contains a glucoside, which he terms *indican;* when indican is decomposed by fermentation, or acted upon by strong acids, it is converted into indigo blue or indigotin, and a peculiar kind of sugar, indiglucin, according to the following formulæ:—

$$\underbrace{C_{52}H_{62}N_2O_{34}}_{\text{Indican.}} + (H_2O)_4 = \underbrace{C_{16}H_{10}N_2O_3}_{\text{Indigo blue.}} + \underbrace{6C_6H_{10}O_6}_{\text{Indiglucin.}}$$

The best indigo is that which has the deepest purple colour, and which assumes, when rubbed with the nail, a bright coppery hue. Its fracture should be homogeneous, compact, fine-grained, and coppery. When reduced to powder it should possess an intense blue colour, and should be so light as to float on water. Indigo should leave only a fine streak when rubbed on a piece of paper. In general, when indigo is in hard dry lumps of a dark colour, or in the form of dust or small pieces, it is frequently adulterated with sand, pulverized slate, and other earthy substances, which fall to the bottom of the vessel when the indigo containing them is thrown on to water. Good indigo leaves only a small quantity of ash on ignition, and when suddenly heated gives off its indigotin in the form of a purplish coloured vapour. A simple and approximative test for indigo consists in drying the sample at 212° Fahr. (100° C.), the loss giving the quantity of hygroscopic water, which should not exceed from 3 to 7 per cent. The dried indigo is next incinerated for the purpose of ascertaining its yield of ash, which in good indigo should not be more than from 7 to 9·5 per cent.

Four pounds of Bengal are equal to five pounds of Guatemala indigo.

There are two methods of preparing solutions of indigo for dyeing :—1. By deoxidizing it, and then dissolving it in alkaline menstrua. 2. By dissolving it in sulphuric acid. The former method is used in preparing the ordinary indigo vat of the dyers.

1. *a*. (COLD VAT).—Take of indigo, in fine powder, 1 lb.; green copperas (clean cryst.), 2½ to 3 lbs.; newly slaked lime,

3½ to 4 lbs.; triturate the powdered indigo with a little water or an alkaline lye, then mix it with some hot water, add the lime, and again well mix, after which stir in the solution of copperas, and agitate the whole thoroughly at intervals for twenty-four hours. A little caustic potassa or soda is frequently added, and a corresponding portion of lime omitted. For use, a portion of this "preparation vat" is ladled into the "dyeing vat," as wanted. After being employed for some time, the vat must be refreshed with a little more copperas and freshly slaked lime, when the sediment must be well stirred up, and the whole thoroughly mixed together. This is the common vat for cotton.

b. (POTASH VAT).—Take indigo, in fine powder, 12lbs.; madder, 8 lbs.; bran, 9 lbs.; potash, 24 lbs.; water at 125° Fahr. (51·5° C.), 120 cubic feet; mix well; at the end of about thirty-six hours add 14 lbs. more potash, and after ten or twelve hours longer, further add 10 lbs. of potash, and rouse the whole up well; as soon as the fermentation and reduction of the indigo are well developed, which generally takes place in about seventy-two hours, add a little freshly slaked lime. This vat dyes very quickly, and the goods lose less of their colour in alkaline and soapy solutions than when dyed in the common vat. It is well adapted for woollen goods. It is worked hot.

c. (WOAD VAT).—As the last, but employing woad instead of madder; the vat is "set" at 160° Fahr. (71° C.), and kept at that temperature until the deoxidation and solution of the indigo has commenced. The last two are also called the "warm vat."

d. (PASTEL VAT).—This is "set" with a variety of woad which grows in France, and which is richer in colouring matter than the plant commonly known as "woad."

e. (SCHÜTZENBERGER and DE LALANDE'S VAT).—It is known that the low stage of oxidation of sulphur obtained on the reduction of sulphurous acid by zinc, dissolves indigo. On this reaction the following proceedings for dyeing and printing with indigo are founded:—To prepare the reducing liquid, a solution of bisulphite of soda at 35°B. is brought into contact with sheet zinc in a closed vessel, of which the liquid should occupy only one-fourth. After the lapse of an hour the zinc is precipitated from the clear liquid by means of milk of lime. It is then diluted or decanted, or filtered with exclusion of air.

The clear liquid is then poured upon the ground indigo, with the addition of the needful soda and lime. One kilo of indigo yields in this manner a very concentrated vat of from 10 to 15 litres. Cotton is dyed cold, and wool with the aid of heat. A vat is filled with water, and a suitable quantity of the above indigo mixture introduced, when the dyeing can be performed at once. The excess of the low sulphur acid dissolves the froth which appears upon the surface. During the process of dyeing, further quantities of indigo can be added as required. Cotton can be rapidly and easily dyed in this manner; and in the case of wool, the dyer escapes the many disadvantages of the hot vat and obtains brighter and clearer shades. To print a fast blue the alkaline solution of the reduced indigo is printed on with an excess of the reducing agent, aged for twelve to twenty-four hours, washed and soaped. In comparison with the old process there is a saving of indigo to the extent of 50 to 60 per cent.; the shades are richer and the impressions sharper. The colour requires no subsequent treatment, and can therefore be printed on simultaneously with most other colours.

f. (GERMAN VAT).—To 2,000 gallons of water heated to 130° Fahr. (54·4° C.), are to be added 20 lbs. of crystals of common carbonate of soda, $2\frac{1}{2}$ pecks of bran, and 12 lbs. of indigo, the mixture being well stirred.

In twelve hours fermentation sets in, bubbles of gas rise,

the liquid acquires a sweet smell, and a green colour; 2 lbs. of slaked lime are next added, with diligent stirring; the vat is again heated and covered over for twelve hours, when a similar quantity of bran, indigo, and soda, with some lime are added.

In about forty-eight hours the vat may be worked; but as the reducing powers of the bran are somewhat feeble, 6 lbs. of molasses are added. Should the fermentation be too energetic, it must be repressed by the addition of lime: if too sluggish, it must be stimulated by the addition of bran and molasses. It is worked hot.

Sulphate of Indigo. *Syn.* SULPHINDYLIC ACID, SULPHINDIGOTIC ACID, SAXONY BLUE, SOLUBLE INDIGO.—This is generally prepared by adding indigo in fine powder 1 part, to Nordhausen sulphuric acid 5 parts, or oil of vitriol 8 parts, contained in a stoneware vessel placed in a tub of very cold water, to prevent the mixture heating. The ingredients are stirred together with a glass rod at short intervals, until the solution is complete, after which the whole is allowed to repose for about forty-eight hours, by which time it becomes a homogeneous pasty mass of an intense blue colour, which in a dull light appears nearly black.

The above preparation, diluted with about twice its weight of soft water, is converted into "Saxony blue."

Wool, silk, linen, and cotton, may each be dyed blue in the indigo vat. The goods after being passed through a weak alkaline solution, are subjected to the action of the vat for about fifteen minutes; they are then freely exposed to the air; the immersion in the vat and the exposure are repeated until the colour becomes sufficiently deep. Woad and madder improve the richness of the dye. Other deoxidizing substances, besides those above mentioned, may be used to effect the solution of the indigo; thus a mixture of caustic soda, grape sugar, indigo, and water, is often employed on the Continent for this purpose; and orpiment, lime, and pearlash are also occasionally used. When properly prepared, the indigo vat may be kept in action for several months by the addition of one or other of its constituents, as required. An excess of either copperas or lime should be avoided.

1. Solution of sulphate of indigo is added to water, as required, and the goods, previously boiled with alum, are then immersed in it, and the boiling and immersion are repeated until the wool becomes sufficiently dyed.

With this, every shade of blue may be dyed, but it is most commonly employed to give a ground to logwood blues. The colouring matter has affinity for wool and silk with or without mordant, but none for cotton. A solution of soluble indigo (sulphindylate of potassa or soda), in water made very slightly acid with sulphuric acid, imparts a very fine blue to cloth, superior in tint to that given by the simple sulphate.

2. Give the goods a mordant of alum, or of acetate of alumina ("red liquor,") then rinse them well, and boil them in a bath of logwood, to which a small quantity of blue vitriol has been added; lastly, rinse and dry.

3. Boil the goods for a short time in a bath of logwood, then add to the liquor tartar and verdigris, in the proportion of 1 oz. of each to every lb. of logwood employed; and again boil for a short time.

Kermes. *Syn. Kermes Grain, Alkermes.*—The dried bodies of the *Coccus ilicis* (LINN.), a small insect which flourishes on the Ilex oak or *Quercus cocciferæ*, a tree growing in the South of France, Spain, Italy, and the Greek Islands. Kermes now only finds use in Spain, Morocco, and Turkey, where it is used for dyeing leather and woollens. The colouring matter yielded by it is nearly the same as that existing in cochineal, but is not so brilliant. It is unacted upon by soap or alkalies.

Lac. *Syn. Lac lake, Indian cochineal.*—The source of this dye, the colouring matter of which is very similar to that found in cochineal, is a species of *Coccus*—the *Coccus laccæ*, a native of India. The insect punctures the branches of certain species of the fig, more especially the *Ficus religiosa indica*, the juice exuding from which in consequence, whilst becoming inspissated, encloses the creature, and at last hardens into a resinous mass around it, which becomes tinged with the colouring principle contained in the insect. By treating this resinous substance, known as *stick lac*, with a weak alkaline solution, and then adding to this a solution of alum, the pigment separates in the form of a lac lake. Lac dye is only suited for woollen or silk goods. It gives scarlet colours like those obtained from cochineal, but of a less brilliant character.

The physical tests of a good lac-dye are that it should be tolerably easily broken by the fingers; that the fracture should exhibit a deep red colour; that it should not have a shining resinous appearance, and that it should evolve a pronounced and peculiar odour. The harder it is, the larger is the amount of shellac, and the smaller the quantity of colouring matter it contains.

It may be tested by putting 5 grains of each sample in a phial, and covering with about 2 fluid drachms of scarlet finishing spirit, and setting aside all the specimens so treated for an hour, after which about an ounce of water is added to each phial, all the phials being then exposed for another hour to a moderate heat. They are then examined as to their respective depths of colour. This examination is all that is necessary when the lacs are used for printing, or for dyeing wool or woollen fabrics.

When used, however, for woollen goods that require to be subsequently hot pressed, the amount of resin contained in the lac must be first ascertained; for if this resin is in excess, the pressing papers will adhere to it in patches, and consequently spoil the goods. To ascertain the amount of resin in a lac, take equal weights of the samples reduced to powder, and place them in small flasks, pouring upon each sample an equal measure of alcohol. The flasks are then loosely stoppered and exposed to a gentle heat, after which the clear solutions are decanted off into capsules, the tares of which have been previously taken. The contents of the capsules are then evaporated to dryness and weighed. The difference in the tare of each capsule will of course represent the amount of shellac in each sample.

A great variety of brands of lac lake are upon the market, but the trade mark is no evidence of excellence.

Messrs. BROOKE, SIMPSON, & SPILLER prepare a lac dye which is said to be superior to that which comes from India. They obtain a lac-lake by dissolving stick lac in weak ammonia, and then adding to the solution chloride of tin.

The colouring matter of stick lac, although not identical with that of cochineal, bears a considerable resemblance to it in its properties.

La Kao.—The *Rhamnus chlorophorus* and *Rhamnus utilis* yield a green dye much used by the Chinese, and formerly by the English dyer, who has latterly abandoned it for the aniline greens.

Logwood. *Syn.* CAMPEACHY WOOD.—This is the heartwood of *Hæmatoxylon campechianum*, a native of the coast of Campeachy, and cultivated in India and the West Indies. *Hæmatoxylin*, the colouring principle of logwood, occurs in brilliant reddish-white or straw-yellow crystals.

When dissolved in water, *hæmatoxylin* forms a colourless solution, which is rendered purple-red by the smallest addition of ammonia. An extract of logwood is very frequently used in dyeing, instead of the wood. In common with other dye extracts, it should be prepared in vacuum pans, and with exclusion, as far as possible, of air, the presence of which acts detrimentally on the colouring matter.

Madder.—The *Rubia tinctorium*, the roots of which yield the madder dye, is a perennial plant growing in the Southern, Central, and Western parts of Europe. Another variety, the *Rubia peregrina*, is largely cultivated in the Levant, and a third, the *Rubiam mungista* or *mungeet*, in India and Japan. The Levant, Indian, and Japanese madders are also sometimes found in the wild state. All the varieties are perennial. The dye-stuff imported under the name of "mungeet" from India, is the reedy stem of a species of *rubia*, and is inferior in tinctorial power to the two other varieties. Large quantities of mungeet are used in Thibet for dyeing the apparel worn by the Llamas. Madder root varies from 4 to 10 inches in length, and is about the thickness of an ordinary goose quill. Deprived of its external brown bark, it presents a yellowish red appearance, and with the exception of the Avignon madder, has a strong smell. The Zealand or Holland madders are distinguished for their marked odour.

The best kind of madder is that grown in the Levant, which occurs in commerce under the name of *lizari* or *alizari*.* The roots of the Levant madder are rather thicker than those of the other varieties, which is due to their being allowed to attain a growth of four or five years, whereas the other roots are used when they are from two to three years old. Madder is sent into the market under the forms of root and powder, the latter being always kept in strong oaken casks, so as to protect it from the action of air and light. European madder root when in powder is technically known as *racine*.

---

* By these terms is understood the entire root of the madder. The term madder is applied to the root when pulverized.

Besides Dutch madder, that from Alsace and Avignon is —or rather, before the extensive employment of the artificial alizarin and purpurin, was—in very large demand, and had a high reputation. Alsatian madder is sent into the market in the state of a very fine powder; and in order to extract its tinctorial principle, it requires boiling a much longer time than the Levant madder. It has a penetrating smell and a bitter taste, and readily absorbs moisture by exposure. It is in best condition after being kept for two years. Avignon madder possesses an agreeable and rather pungent odour. It is met with in the condition of a very fine powder. It absorbs moisture less readily than the other species. That which has been kept in casks for a year is to be preferred for use. It keeps well and undergoes little if any fermentation. The finest quality of ground madder is called *crop*, or *grappe;* next come *ombro* and *gamene*, and lastly *mull*, which consists of the refuse and dust from the madder grinding rooms.

Flowers of madder, or *fleur de garance*, as the preparation is called by the French dyers, is obtained by infusing 1 part of madder in fine powder in 8 or 10 parts water, and setting up fermentation in the liquid, whereby the large amount of sugar* which the root contains is removed. The residue is then thoroughly washed with warm and afterwards with cold water. The residue being next freed from water by means of hydraulic pressure, is carefully dried, and after being again reduced to powder is ready for use. It is used in the same manner as madder, except that in the dye-beck it is subjected to a lower temperature. By the above treatment the pectous substances of the madder root, which

---

* The fermented liquid, being submitted to distillation, yields a spirit, which is employed for technical purposes.

would otherwise become insoluble during the process of dyeing, are eliminated.

Garancine is obtained by first moistening madder root, reduced to fine powder, with water, and next adding ½ part of sulphuric acid, diluted with double as much water. The mixture is next heated by steam for an hour, and the acid removed from the magma by well washing this latter in water. The resulting garancine, next submitted to hydraulic pressure, whereby the water is removed, is then dried and finally ground to a very fine powder, which amounts to about 25 per cent. of the madder root operated upon. The sulphuric acid destroys much of the woody fibre and other substances which interfere with the dyeing properties of the madder. The tinctorial value of garancine is three or four times that of madder.

When the fluids of the beck left after dyeing with madder root, are strained from the solid residue, and this is treated with half its weight of sulphuric acid, and the resulting mass, in a manner similar to that followed in preparing garancine, a product is obtained, which after drying, goes by the name of *garanceux*. *Garanceux* is mostly used for producing *sad* colours. It is inferior in tinctorial power to *garancine*.

Madder Extracts are made by treating madder root with boiling water, collecting the precipitates which form as the infusion cools, mixing them with gum or starch, and then adding acetate of alumina or iron. Extracts so prepared form mordanted dyes, which are available for direct application in calico printing.

The consumption of madder, and the various tinctorial preparations obtained from it, have, owing to the recent introduction of artificial alizarin and purpurin, greatly declined in England.

Madder root contains several distinct principles, such

as *madder red* or *alizarin*, *madder purple* or *purpurin*,[*] *madder orange* or *rubiacin*, *madder yellow* or *xanthin*.

The researches of SCHUNCK have shown that alizarin, the colouring principle of madder root, is derived from a glucoside which he terms *rubian*, which, under the influence of acids and alkalies, and of a peculiar nitrogenized ferment known as *erythrozym*, splits up into alizarin and other colouring matter, and a fermentable sugar.

The following equation, according to GERHARDT, represents the reaction that takes place :—

$$\underbrace{C_{16}H_{16}O_9}_{\text{Rubian.}} + H_2O = \underbrace{C_{10}H_6O_3}_{\text{Alizarin.}} + \underbrace{C_6H_{12}O_6}_{\text{Glucose.}}$$

The chief mordants used in madder dyeing and calico-printing, are the acetates or pyrolignites of alumina and iron, the first known as " red liquor ;" the second as " black" or " iron liquor."

Formulæ for Red Liquor :—

1. URE. *Standard Red Liquor.*
    Alum . . . . 20 lbs.
    Sugar of lead . . . . . . 12½ ,,
    Boiling water . . . . . . 5 gallons.
   Stir till dissolved ; let settle and draw off the clear.

2. KŒCHLIN.
    Alum . . . . 11 kilos.
    Acetate of lead . . . . . . 82·5 grammes.
    Boiling water . . . . . . 32 litres.

According to KŒCHLIN, this mordant produces the deepest tints with nearly all tinctorial substances.

---

[*] Purpurin occurs in crystalline red needles, insoluble in cold, but soluble in hot water, and in alcohol, ether, and solutions of the alkalies. Its formula is $C_6H_6O_3$.

3. *Five-fourths Mordant.*

Alum	625 grammes.
Acetate of lead	450 ,,
Boiling water	2 litres.

The following, which are French formulæ for red liquor, have a high reputation:—

4.

	A.	B.	C.
Alum	16 kilos.	80 kilos.	10 kilos.
Acetate of lead	12 ,,	8·5 ,,	10 ,,
Boiling water	62 ,,	60·0 ,,	20 ,,
Extract of Lima wood at 20° Baumé.	2 ,, 30° Twaddle.	4·0 ,, 4° Twaddle.	(at 3°)

It is customary with English dyers to make up the above without the peachwood liquor, and to add it when the colour is being prepared for printing.

5. *For Garancine* (MUSPRATT).

At 11° Baumé = 1·083 sp. gr. = 15° Twaddle.

Alum	25 kilos.
Acetate of lead	19 ,,
Water	80 litres.

6. *Strong Mordant.*

At 11° Baumé.

Alum	2½ kilos.
Acetate of lead	2 ,,
Water	6·3 litres.

The German dyers frequently prepare their red liquor from basic alum, which they dissolve in acetic acid, the basic alum being first obtained by treating alum with carbonate of soda. Acetate of lime being a cheaper commodity than lead acetate, is frequently substituted for it. The chlorides of ammonium, sodium, and zinc are added to red liquor, since they prevent the too rapid drying of the acetate on the fibre. Tartrate of alumina, which is formed when cream of tartar is mixed with alum, is found to be superior to the acetate for dyeing woollen goods.

"Iron liquor" or "black liquor" is made as follows :—

1. Copperas, 300 lbs., dissolved in 175 gallons of hot water; to the solution is added 57 gallons of acetate of lime liquor at 16° Twaddle.

2. Copperas 32 lbs.; pyroligneous acid at 7° Twaddle; acetate of lime liquor, at 24° Twaddle, 10 gallons.

Iron liquor at about 6° Twaddle, when properly thickened, gives black with madder. From 4° Twaddle downwards to a very diluted state, it gives various shades of purple or lilac; mixed with red liquor, it gives chocolates.

The purified pyroligneous acid should be employed in the preparation of these mordants, as the tarry matters contained in the crude acid act injuriously. Starch is the best thickener for iron liquor.

Madder is adulterated with powdered brick, ochre, yellow sand, yellow clay, oak sawdust, mahogany sawdust, fustic, and various dye woods. The presence of any foreign mineral matter may be detected by the amount of ash yielded on incineration. Good madder should yield from 9 to 11 per cent. of ash.

The best method of estimating the tinctorial value of a sample of madder is to compare its dyeing power with a specimen of known good quality, which is carried out by the following process, given by Dr. CALVERT :—Place 12 grains of each sample in pans of copper or block-tin with a quart of water. The pans are placed in a water-bath, heated by means of a jet of steam. A piece of calico, mordanted with red, purple, and chocolate mordants, which cover about three-fourths of the surface of the cloth, is placed in each pan.

It is important that each strip taken should be about

3 inches in breadth, and its length equal to one-half the breadth of the calico (26 inches). The swatches are placed in the pans whilst cold, steam is then turned on, and the temperature is gradually raised during an hour and a half to 180° Fahr. (82° C.), and then for half an hour kept as near the boiling-point as possible. During the whole time of the operation the pieces should be constantly and carefully lifted out of the dyeing liquor, either with a glass-rod, or better still, by a mechanical arrangement.

When the dyeing is completed, the pieces are thoroughly washed with pure water, and the brilliancy and intensity of shade carefully compared. If the samples under trial are found to be weaker than the standard, the dyeing operation is repeated, adding such a quantity of the inferior madder as will bring up the colour to the same intensity as the standard. The values are in inverse ratio to the quantities taken. A further trial is, however, necessary to arrive at a correct conclusion as to the value. The dyed pieces are divided into two parts, one of which is kept for comparison, whilst the other is submitted to a light soaping; three or four grains of soap to a quart of water being sufficient for the surfaces above given. They are carefully heated in this solution for a quarter of an hour, the temperature being kept at 180° Fahr. (82° C.). They are then washed and dried, and the tints again compared. The first operation gives the total amount of colour, the second removes any colouring matter of the dye woods which may have been used for the purposes of adulteration.

Murexid. *Syn.* PURPURATE OF AMMONIA.—By adding to a solution of alloxan and alloxantin a solution of carbonate of ammonia, magnificent iridescent crystals of a beautiful reddish purple by transmitted, and of an equally beautiful green colour by reflected, light are obtained. Some years back this substance, known as " murexid," was extensively used in dyeing, but it is now almost, if not entirely, superseded by the coal-tar colours.

Peachwood. *Syn.* ST. MARTHA'S WOOD.—The source of this dye-stuff is the *Cæsalpinia echinata*. The best is imported from Nicaragua, and an inferior kind from Sierra Nevada.

Peachwood, sapanwood, and Lima wood, all give very similar shades of colour.

Persian Berries. *Syn.* YELLOW BERRIES. FRENCH BERRIES. AVIGNON BERRIES. BERRIES.—The berries or fruit of different species of *Rhamnus*, growing in Persia, the Levant, Southern France, and Hungary. The Persian berries, which are the most valuable, are stated to be the product of the *Rhamnus amygdalinus*. Two varieties of these berries are met with in commerce—the larger, a bright olive-coloured berry, which is gathered before being ripe, and a smaller, shrivelled, deep brown kind, which are not removed from the branches until some time after they have reached maturity. The yellow colouring matter of these berries is *chryso-rhamnin* which occurs in golden yellow crystals. These, when boiled in water, are resolved into *xantho-rhamnin*, a substance of an olive yellow colour. According to BOLLEY, *chryso-rhamnin* is identical with *quercetin*. Persian berries are chiefly used for dyeing morocco leather, yellow. Cloth, previously mordanted with alum, tartar, or protochloride of tin is also dyed yellow, whilst with sulphate of copper they give an olive, and with sulphate of iron an olive-green colour.

Purree. *Syn.* INDIAN YELLOW.—This yellow dye-stuff, imported from China and India, is of doubtful origin, and is more largely used by artists than by dyers. Some writers suppose it to be derived from an animal source. According to the researches of STENHOUSE and ERDMANN, purree chiefly consists of purreic acid, a strongly tinctorial substance, combined with magnesium.

Quercitron Bark. *Syn.* BARK.—This yellow dye material is the inner bark of the *Quercus tinctoria, Quercus nigra*, or *Quercus citrina*, a species of oak, a native of America, growing more particularly in Pennsylvania, Carolina and Georgia.

Quercitrin, the pigment of quercitron bark, is a neutral substance which, when treated with dilute acids, yields *quercetin*, a substance of a lemon-yellow colour, which occurs in commerce under the name of *flavine*. With picric acid, quercitron bark affords a magnificent yellow dye.

Rhubarb.—The root of common rhubarb contains a yellow colouring principle, termed *rhein* or *chrysophanic acid*, which is soluble in boiling achohol and ether, from which it crystallizes in golden yellow crystals of a metallic appearance. With alkalies it produces a reddish-brown coloured liquid.

Sandal Wood.—This is the wood of the *Pterocarpus santalinus*, an Indian tree. It is imported in logs, which are of a deep red colour externally, and a bright red internally. The colouring principle of sandal wood resides in a resinoid body named *santalin*, which appears to be the oxidation product of a colourless body named *santal*. *Santalin* is also found in barwood, the source of which is *Baphia nitida*, an African tree. According to BANCROFT, if wool be mordanted with alum and tartar, and then dyed in a bath of sandal wood and sumach, it takes a reddish-yellow colour.

Safflower. *Syn.* BASTARD SAFFRON. DYERS' SAFFRON.— The dye-stuff to which this name is given, consists of the dried florets of *Carthamus tinctorius*, a thistle-like plant, cultivated in Spain, Egypt, the Levant, and in some parts of Germany. Safflower contains two colouring principles— a yellow and a red one. The latter, which is the tinctorial agent of the plant, is called *carthamin*. The yellow is removed by water and is rejected. The red is easily dissolved out from the florets by weak solutions of the carbonated alkalies, and is again precipitated on the addition of an acid. Safflower is employed for dyeing silk, to which it imparts a brilliant but very fugitive colour.

Saffron.—Saffron is the prepared stigmata or stigmas of the *Crocus sativus*, or saffron crocus. There are two principal varieties of saffron known in commerce : hay saffron, and cake saffron. Hay saffron consists of the stigmas with part of the styles, carefully separated from the other part of the flowers, and then dried by a very gentle heat. Cake saffron is the last kind compressed into a cake, after it has been softened by the fire and afterwards dried. Saffron is very largely and constantly adulterated. The chief sophisticants are safflower, marigold, and carbonate of lime. Saffron owes its colour to *crocin*, a glucoside.

Turmeric.—This dye-stuff is the dried rhizome or underground stem of the *Curcuma longa*, and the *Curcuma rotunda*, a plant growing in India, Java, and Ceylon. The finest kinds come from the latter island. *Curcumin*, the colouring principle of turmeric occurs as a brownish yellow mass. It dyes cotton without a mordant. It gives a golden yellow to wool, and an orange tinge to scarlet. It is, however, a very fugitive dye.

Weld.—Weld consists of the dried herbs and stems of the *Reseda luteola*, a native of the South of France. *Luteolin*, the colouring matter of weld, is a substance of considerable durability, and when sublimed condenses in yellow needles. The decoction of weld imparts a rich yellow to goods mordanted with alum, tartar, or chloride of tin.

Woad. *Syn.* DYER'S WOAD ; PASTEL, Fr.—The *Isatis tinctoria*. To prepare them for the dyer, the leaves of the plant are partially dried and ground to a paste, which is made into balls ; these are placed in heaps, and occasionally sprinkled with water, to promote the fermentation ; when this is finished, the woad is allowed to fall down into lumps, which are afterwards reground and made into cakes for sale. The woad thus prepared is mixed with boiling water, and allowed to stand for some hours in a closed vessel, about 1-20th its weight of newly slaked lime being added to it. The mixture is then digested at a moderate temperature, and stirred every three or four hours, when a new fermentation begins; a blue froth rises to the surface, and the liquor, though it appears itself of a reddish colour, dyes woollens of a green which, like the green from indigo, changes in the air to a blue. It is said this process does not succeed well on the small scale. Woad is now mostly used in combination with indigo. Fifty lbs. of woad are reckoned equal to 1 lb. of indigo.

By increasing the proportion of alum or red-liquor, the colour verges on purple; and by employing a little acetate of iron, or green copperas, the darker shades of blue are produced. Verdigris, blue vitriol, and alkalies, turn it more on the blue; whilst a mordant of tin imparts a violet cast.

Coal-Tar Colours.—The old synonym of . aniline colours is hardly now applicable, as these only form a portion of the colours obtained from TAR. Coal-tar consists of the oily fluid obtained in the destructive distillation of coal, during the manufacture of ordinary illuminating gas, and collected in a tank from the hydraulic main and condensers. The composition of coal-tar is highly complex, the most important constituents, being, however, benzol, toluol, naphthalene, anthracene and carbolic acid. Naphtha, which is one of a series of homologous hydrocarbons obtained by distilling coal-tar, yields, by rectification, between 180° and 250° Fahr. (82° and 121° C.), an almost colourless liquid, the benzol of commerce. By the action of a mixture of nitric and sulphuric acids on benzol, nitro-benzol, a heavy oily liquid with an odour of oil of bitter almonds, is obtained. In commerce this substance is made in large cast-iron pots, fitted with tight covers, and provided with stirrers worked by steam power. By means of pipes the reagents are admitted, and the nitrous fumes are carried off, while the nitro-benzol and the spent reagents are drawn off from the bottom. The entire charge of benzol is first placed in the vessels, and the mixed acids are, as the reaction is very energetic, cautiously run in, the whole being well stirred throughout. When finished, the contents are drawn off, and the nitro-benzol collected, washed with water, and finally washed with a weak solution of soda. Nitro-benzol is converted into aniline in a similar apparatus, only provided with the means of admitting a current of superheated steam, and condensing the aniline as it distils over. In the vessel iron borings are placed, and acetic acid and nitro-benzol cautiously run in as the reduction is violent, stirring well all the time. A current of superheated steam is passed through, and the aniline collected as it distils over, as a pale, sherry-coloured, oily liquid, boiling at 360° Fahr. (182° Cent.), and of sp. gr. 1·02.

Aniline. C_6H_7N. *Syn.* PHENYLAMINE.—A peculiar volatile organic base first noticed by UNVERDORBEN in empyreumatic bone-oil, and afterwards obtained by RUNGE from coal-tar, and by FRITZSCHE, ZININ, A. W. HOFMANN, and others, as a product of various reactions, processes, and decompositions, particularly those attending the destructive distillation of nitrogenous organic bodies. Aniline is now invariably obtained, on the large scale, indirectly from coal-tar, from the naphtha, or, more correctly, from the nitro-benzol, of which this is the source. The following are the leading commercial and experimental processes:—

1. From COAL-TAR.—The aniline present in coal-tar may be obtained by washing the crude naphtha with dilute hydrochloric acid, the clear portion of the liquid (containing the hydrochlorates of the bases present) is then decanted and

carefully evaporated over an open fire until acrid fumes begin to be disengaged, when it is again decanted or filtered; the clear liquor, or filtrate, is next treated with caustic soda or milk of lime in excess, by which the bases are liberated under the form of a brownish oil; the whole of the resulting mixture is now submitted to distillation, the portion which passes over at or about 360° Fahr. (182° Cent.), and which consists chiefly of crude aniline, being collected separately; the product is purified by rectification at the same temperature, and, lastly, by fresh treatment with hydrochloric acid, and careful distillation with excess of soda, or milk of lime, as before.

2. From NITRO-BENZOL:—*a.* (ZININ.) An alcoholic solution of nitro-benzol, after saturation with ammonia, is treated with sulphuretted hydrogen, until, after some hours, a precipitation of sulphur takes place; the brown liquid is then repeatedly saturated with fresh sulphuretted hydrogen, until no more sulphur separates, the reaction being aided by occasionally heating or distilling the mixture; an excess of acid is next added, and, after filtering the liquid, and the removal of the alcohol and unaltered nitro-benzol by ebullition or distillation, the residue is lastly distilled with caustic potash, in excess. The aniline found in the receiver may be rendered pure by forming it into oxalate of aniline, repeatedly crystallizing the salt from alcohol, and finally distilling it with excess of caustic soda as before.

The following is a cheaper and more convenient process, and is the only method used on a large scale:—

b. (M. BÉCHAMPS). From NITRO-BENZOL by distillation with a mixture of iron-filings and acetic acid (the acetic acid is now more economically replaced by hydrochloric acid or a solution of chloride of iron).

The liquor found in the receiver consists of aniline and water, from which the first, forming the lower portion, is obtained, after sufficient repose in a separator. A very spacious retort must be employed in the process, as the mass swells up violently; and it must be connected with the receiver, by means of a condenser, kept in good action by a sufficient flow of cold water.

The apparatus for carrying out BÉCHAMP's method was devised by NICHOLSON, and is exhibited in the subjoined plate.

"It consists essentially of a cast-iron cylinder (A) of 10 hectolitres (220 gallons) cubic capacity. A stout iron tube is

FIG. 5.

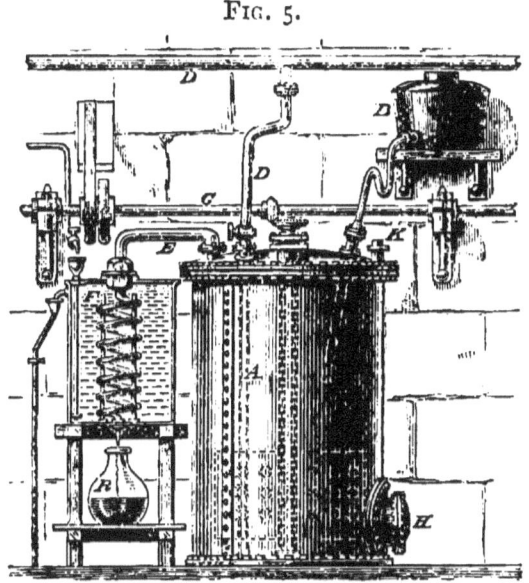

fitted to this vessel, reaching nearly to the bottom of the cylinder. The upper part of this tube is connected with the machinery G, while the surface of the tube is fitted with steel projections. The tube serves to admit steam, as well as acting as a stirring apparatus. Sometimes, instead of this tube, a solid iron axle is employed, and in this case there is a separate steampipe, D. Through the opening at K the materials for making aniline are put into the apparatus, while the volatile products are carried off through E. H serves for emptying and cleaning the apparatus. The S-shaped tube connected with the vessel B acts as a safety valve. When it is intended to work with this apparatus there is poured into it through K, 10 kilos of acetic acid at 8° B. (sp. gr. 1·060), previously diluted with six times its weight of water; next there are added 30 kilos of iron filings, or cast iron borings, and 125 kilos of nitro-benzol, and immediately after the stirring apparatus is set in motion. The reaction ensues directly, and is attended by a considerable evolution of heat and vapours. Gradually more iron is added until the quantity amounts to 180 kilos. The escaping vapours are condensed in F, and the liquid condensed in R is from time to time poured back into the cylinder A. The reduction is finished after a few hours."*

3. From INDIGO.—Powdered indigo is added to a boiling and highly concentrated solution of caustic potash, as long as it dissolves and hydrogen is liberated, the resulting brownish-red liquid is evaporated to dryness, and the residuum is submitted to destructive distillation in a retort.—*Prod.* 18 to 20 per cent. of the indigo employed.

4. By fusing, with proper precautions, a mixture of isatin and hydrate of potassium (both in powder). A retort connected with a well-cooled receiver, is employed as the apparatus. The interest attaching to these two methods arises from the fact that the aniline thus obtained is absolutely free from toluidine, which is always present in that prepared from coal-tar benzol.

5. From anthranilic acid mixed with powdered glass or sand, and rapidly heated in a retort.

---

* WAGNER's "Chemical Technology," edited by W. CROOKES, F.R.S.

6. By treating an alcoholic solution of benzine with a little zinc and hydrochloric acid.

7. By heating phenyl-alcohol with ammonia in sealed tubes.

Many other reducing agents have been proposed for the conversion of nitro-benzol into aniline, such as arsenite of sodium, powdered zinc, &c., but on the large scale they have all been found inferior to the process of BÉCHAMP. KREMER's process consists in heating one part of nitro-benzol in a proper apparatus with five of water and two and a half of zinc dust. When the reaction is completed, the aniline, amounting to about 65 per cent. of the weight of the benzol, is distilled off in a current of steam.

Aniline is a thin, oily, colourless liquid, with a faintly vinous odour, and a hot and aromatic taste, miscible in all proportions with alcohol and ether, very slightly soluble in water, neutral to ordinary test-paper, but exhibiting an alkaline reaction to dahlia-petal infusion and paper. It dissolves camphor, sulphur, and phosphorus, and coagulates albumen, possesses a high refractive power, and precipitates the oxides of iron, zinc, and alumina, from solutions of their salts, and neutralizes the acids like ammonia. With the acids it forms numerous crystallizable compounds of great beauty, which are easily formed, and are precisely analogous to the corresponding salts of ammonia. These, on exposure to the air, acquire a rose colour, in many cases gradually passing into brown.

Tests.—1. Chromic acid gives a deep greenish or bluish-black precipitate with aniline and its salts. 2. Hypochlorite of lime strikes an extremely beautiful violet colour, which is soon destroyed. 3. The addition of two or three drops of nitric acid to anhydrous aniline produces a fine blue colour, which, on the application of heat, passes into yellow, and a violent reaction ensues, sometimes followed by explosion. 4. With bichloride of platinum it yields a double salt, the platino-chloride of aniline, corresponding to the corresponding salt of ammonia. These reactions distinguish it from all other substances.

Commercial aniline is a mixture consisting in great part of aniline, paratoluidine, and orthotoluidine in variable proportions. In addition it contains small amounts of metatoluidine, nitro-benzol, odorine, &c., but for all practical purposes it may be regarded as a mixture of aniline and toluidine. As it is obtained from a portion of the light naphtha, boiling between certain temperatures, it will vary according to the naphtha from which it is made.

In order to distinguish between various samples of commercial aniline, REIMANN submits them to fractional distillation and compares the results. He places 100 c.c. of the sample to be tested in a retort fitted with a thermometer and heated by means of an oil-bath. The liquid, as it distils, is received in a narrow graduated cylinder, and the amount that passes over between every 5° Cent. (41° Fahr.) is noted.

In order to obtain standards for comparison he first distilled a sample of light aniline, then one of heavy aniline;

CENTIGRADE.	Light 100 Heavy 0	90 10	85 15	80 20	75 25	60 40	50 50	25 75	0 100
Below 180°	8½	7	2½	5½	7	...	7	5½	...
180°—185°	54	50	29½	22	5½	7	4½	2½	2
185°—190°	34	34	56½	55½	55½	37	7½	4½	1½
190°—195°	...	5	7½	8½	15	33	42	17	S
195°—200°	...	...	...	...	9	...	19	36	18
200°—205°	...	...	...	...	4½	16	10	16	39
205°—210°	...	...	...	...	..	...	3½	8	19
210°—215°	...	...	...	..	...	...	..	4½	7
Residue .	3½	4	4	8½	3½	7	6½	5	5½

and afterwards mixtures of the two in varying proportions. In the foregoing table the results are given.

In examining commercial aniline it is usual to determine the amount of insoluble oil which separates on agitating a sample with dilute hydrochloric acid. For a detailed account of the methods of preparing aniline commercially, and of the dyes obtained therefrom, *see* "Dictionnaire de Chimie," par A. WURTZ.

Aniline Black is formed as a by-product in the manufacture of all colours which are produced by the action of powerful oxidizers on aniline. In the case of mauve, for instance, the black by-product amounts to as much as thirty times the weight of the pure colour. There are so many methods of preparing it that the two following may be mentioned as examples:—1. "Dissolve 20 parts of potassium chlorate, 40 parts of sulphate of copper, 16 parts of chloride of ammonium, and 40 parts of aniline hydrochlorate, in 500 parts of water, warming the liquid to about 60°, and then removing it from the water-bath. In about three minutes the solution froths up and gives off vapours which strongly attack the breathing organs. If the mass does not become quite black after the lapse of a few hours it is again heated to 60° Fahr. (15·5° C.), and then exposed in an open place for a day or two, and afterwards carefully washed out till no salts are found in the filtrate. For use in printing, the black paste is mixed with a somewhat large quantity of albumen, and the goods after printing are strongly steamed. The paste can be pressed into moulds, and used as a substitute for Indian ink."[*]

2. "Mix equal weights of aniline (containing toluidine), hydrochloric acid, and potassium chlorate, with a minute quantity of cupric chloride and a sufficient quantity of water, and leave the mixture to evaporate spontaneously, when a black powder will be obtained."[†]

---

[*] A. MÜLLER.   [†] RHEINECK.

Aniline Blue, or Bleu de Lyons.—This dye is prepared by heating a mixture of magenta, acetate of sodium, and aniline, in iron pots, provided with stirrers, &c., in an oil bath, to 370° Fahr. (188° Cent.) When a good blue has been obtained, the heat is removed, and the thick treacle-like fluid purified. This is effected, for the commoner varieties, by treating the crude product with hydrochloric acid, to dissolve out the excess of aniline, and the various red and purple impurities.

The better qualities, however, termed Opal Blues, are prepared by heating purified magenta base with pure aniline and a certain quantity of acetic acid (this acid is sometimes replaced by benzoic or naphthoic acid). The crude product thus obtained is mixed with methylated spirit, and poured into water acidulated with hydrochloric acid. The colour that is precipitated is then collected on filters, washed and dried. This blue, like magenta, is a salt of a colourless base, which has been named Triphenylrosaniline, $C_{38}H_{31}N_3$, or $C_{20}H_{16}(C_6H_5)_3N_3$. Aniline blue, or Lyons blue, is sent into the market either as a coarse powder of a coppery lustre, or in alcoholic solution, as it is insoluble in water.

Mr. NICHOLSON, by treating Lyons blue in the same manner as indigo is converted into sulpindigotic acid, succeeded in rendering it soluble, and thus prepared the colours known as *Soluble Blues*.

Nicholson's Blue is obtained by digesting triphenylrosaniline or monosulphonic acid* with a quantity of soda-lye not quite sufficient for saturation, filtering the solution and evaporating. It is dried at 100° Fahr. (38° C.) Wool dipped into a hot aqueous solution of Nicholson's blue,

---

* This acid is made by dissolving triphenylrosaniline hydrochloride in strong sulphuric acid, and heating the solution for five or six hours. On the addition of water, the acid is obtained as a dark blue precipitate, and dried at 100° Fahr. (38° C.).

especially if borax, or water-glass be added, extracts it in a colourless state, and holds it so fast that it cannot be washed out with water, but on dipping the wool thus prepared into an acid, the salt is decomposed, and the colouring matter is set free.

Mauve, the first-discovered coal-tar or aniline colour, was obtained by Mr. PERKIN during some experiments directed towards the artificial formation of quinine, and was also first practically manufactured by Mr. PERKIN. Commercially, mauve is made as follows : Aniline and sulphuric acid, in proper proportions, are dissolved in water in a vat by aid of heat, and when cold a solution of bichromate of potassium is added, and the whole allowed to stand for a day or two, when a black precipitate is obtained, which, after being collected on shallow filters, is washed and well dried. This black resinous substance is digested with dilute methylated spirit in a suitable apparatus to dissolve out the mauve, and the spirit is distilled off. The mauve is precipitated from the aqueous solution left behind by hydrate of sodium, and after washing drained to a paste.

The amount of mauve thus obtained is small in comparison with the raw material, coal tar; since 100 lbs. of coal yield 10 lbs. 12 oz. of coal tar; $8\frac{1}{2}$ oz. of mineral naphtha, $2\frac{3}{4}$ oz. of benzol; $4\frac{1}{4}$ oz. of nitro-benzol, $2\frac{1}{4}$ oz. of aniline, and $\frac{1}{4}$ oz. of mauve. Mauve is usually sent into the market in paste or solution, the expense of the crystals being heavy, and offering no corresponding advantages.

Other salts than the bichromate of potassium have been employed to convert aniline into mauve, such as chloride of copper, permanganate of potassium, &c.; but experience has shown none to possess the same advantages as the bichromate of potassium.

Mauveine, the organic base of mauve or aniline purple, is a black crystalline powder, yielding a dull violet solution.

The moment, however, mauveine is brought into contact with an acid, it turns a magnificent purple colour. The salts of mauveine form beautiful crystals, possessing a splendid green metallic lustre.

Aniline Brown. *Syn.* **Habana Brown.**—DE LAIRE prepares this pigment by heating to a temperature of 464° Fahr. (240° Cent.), a mixture of aniline violet and aniline blue with hydrochlorate of aniline. The product thus obtained dissolves in water, acids, and alcohol, and can be used at once for dyeing purposes.

BISMARK'S BROWN is made by fusing fuchsin with hydrochlorate of aniline.

Dahlia.—This is prepared from mauve and iodide of ethyl, in the same manner as the HOFMANN violets, and is a purple-red violet. It is a good colour, but the expense precludes its general use.

Aniline Pink, a very clear and brilliant dye, is obtained by the action of binoxide of lead and acetic acid on mauve; this colour is also formed as a by-product in the manufacture of the last-mentioned dye, but on account of its being dissolved in so large a quantity of liquor it is lost. WILLM, in 1861, obtained another red colouring matter, which he got by boiling an acetic solution of mauveine with binoxide of barium.

Aniline Grey.—Mauveine paste is dissolved in rather more than its own weight of strong sulphuric acid, and two-thirds of its weight of aldehyde is then added. After being stirred, the mixture is allowed to stand four or five hours; it is then poured into water, and the solution, after it has been filtered, yields the grey colouring matter on the addition of salt.

SCHEURER-KESTNER obtained a yellow dye by the action of tin on a solution of mauveine, in hydrochloric acid.

Aniline Blue for Printing.—BLUMER-ZWEEFEL gives the following process for preparing this colour:—" Mix

100 parts of starch with 1,000 parts of water, and add to it, while warm, 40 parts of potassium chlorate, 3 to 4 parts of ferrous sulphate, and 10 parts of sal ammoniac. The well-mixed paste, when quite cold, is mixed with 70 parts of aniline hydrochloride, or an equivalent quantity of tartrate, and immediately used. The printed goods are oxidized, then passed through warm or faintly alkaline water, whereby the blue colour is developed."

Another colouring matter, called Paris blue or Bleu de Paris, was obtained by heating stannic chloride with aniline for thirty hours at a temperature of 356° Fahr. (180° C.). It is a fine pure blue, soluble in water, and crystallizing in large blue needles with a coppery lustre.

Aniline Green.—When treated with chlorate of potassium, to which a quantity of hydrochloric acid has been added, aniline assumes a rich indigo-blue colour. The same result occurs if the aniline be treated with a solution of chlorous acid. Similar blues have been obtained by Messrs. CRACE CALVERT, LOWE & CLIFT. Most of these blues possess the property, when subjected to the action of acids, of acquiring a green tint, called Emeraldine. Dr. CALVERT obtained this colour directly upon cloth, by printing with a mixture of an aniline salt and chlorate of potassium, and allowing it to dry. In about twelve hours the green colour is developed. This colour may be converted into blue by being passed through a hot dilute alkaline solution, or through a bath of boiling soap.

Saffranine.—This dye-stuff is of a bright red-rose colour. MENÉ says it may be prepared commercially by treatment of heavy aniline oils successively with nitrous and arsenic acids; or 2 parts of the aniline may be heated with 1 of arsenic acid, and 1 of an alkaline nitrite for a short time, to 200° or 212° Fahr. (95°–100° C.). The product is extracted with boiling water neutralized with an akali, filtered, and the colour thrown down by common salt.

Magenta. *Syn.* ANILINE RED, ROSEINE, FUCHSINE, AZALEINE, SOLFERINO, TYRALINE.—Various processes have been proposed and patented for the preparation of this commercially important coal-tar colour. Amongst these processes are—

1. GERBER-KELLER's, patented in France, October 29, 1859. By this the aniline is treated with mercuric nitrate.

2. LAUTH & DEPOUILLY used nitric acid.

3. MEDLOCK (patent dated January, 1860), NICHOLSON, and GIRARD & DE LAIRE, all in 1860, separately patented the use of arsenic acid. This process, being the one now almost exclusively employed, is thus described in CRACE CALVERT's work, "Dyeing and Calico-Printing," edited by Messrs. STENHOUSE & GROVES : —" The manufacture of magenta, as it is now conducted in the large colour works, is a comparatively simple process, the apparatus employed consisting of a large cast-iron pot set in a furnace, provided with means for carefully regulating the heat. It is furnished with a stirrer, which can be worked by hand or by mechanical means, the gearing for the stirrer being fixed to the lid, so that by means of a crane the lid may be removed, together with the stirrer and gearing. There is also a bent tube passing through the lid for the exit of the vapours, which can be easily connected or disconnected with a worm at pleasure ; lastly, there are large openings at the bottom of the pot, closed by suitable stoppers, so that the charge can be removed with facility as soon as the reaction is complete. Into this apparatus, which is capable of holding about 500 gallons, a charge of 2,740 lbs. of a concentrated solution of arsenic acid, containing 72 per cent. of the anhydrous acid, is introduced, together with 1,600 lbs. of commercial aniline. The aniline selected for this purpose should contain about 25 per cent. of toluidine.

"After the materials have been thoroughly mixed by the stirrer the fire is lighted, and the temperature gradually raised to about 360° Fahr. (182° C.). In a short time water begins to distil, then aniline makes its appearance along with the water, and, lastly, aniline alone comes over, which is nearly pure, containing, as it does, but a very small percentage of toluidine. The operation usually lasts about eight or ten hours, during which time about 170 gallons of liquid pass over, and are condensed in the worm attached to the apparatus; of this about 150 lbs. are aniline. The temperature should not exceed 380° Fahr. (193° C.) at any period during the operation. When this is complete, steam is blown in through a tube, in order to sweep out the last traces of the free aniline, and boiling water is gradually introduced in quantity sufficient to convert the contents into a homogenous liquid. When this occurs the liquid is run out of the openings at the bottom, into cisterns provided with agitators; here more boiling water is added, in the proportion of 300 gallons to every 600 lbs. of crude magenta, and also 6 lbs. of hydrochloric acid. The mass is then boiled for four or five hours by means of steam pipes, the agitators being kept in constant motion. The solution of hydrochloride, arsenite, and arseniate of rosaniline thus obtained is filtered through woollen cloth, and 720 lbs. of common salt added to the liquid (which is kept boiling) for each 600 lbs. of crude magenta. By this means the whole of the rosaniline is converted into hydrochloride, which, being nearly insoluble in the strong solution of arseniate and arsenite of sodium produced in the double decomposition, separates and rises to the surface; a further quantity is deposited from the saline solution on allowing it to cool and stand for some time. In order to purify the crude rosaniline hydrochloride it is washed with a small quantity of water, redissolved in boiling water slightly acidulated with hydrochloric acid, filtered, and allowed to crystalize."

In the treatment of aniline with arsenic acid, violet and blue dyes are also formed. The production of such has been patented by GIRARD & DE LAIRE.

4. LAURENT & CASTHÉLAZ have obtained aniline red direct from benzol, without the preliminary isolation of aniline. Nitro-benzol is treated with twice its weight of iron finely divided, and half its weight of concentrated hydrochloric acid. The colouring matter obtained by this process is said to be inferior in beauty to that procured from aniline.

5. RENARD BROTHERS include in their patent the ebullition of aniline with stannous, stannic, mercurous, and mercuric sulphates, with ferric and uranic nitrates and nitrate of silver, and with stannic and mercuric bromides.

6. DALE & CARO's (patent dated 1860) consists in the treatment of aniline or hydrochlorate of aniline, with nitrate of lead.

7. SMITH claims the ebullition of aniline with perchloride of antimony, or the action of antimonic acid, peroxide of bismuth, stannic, ferric, mercuric, and cupric oxides, upon hydrochlorate or sulphate of aniline, at the temperature of 180°.

COUPIER's process for the manufacture of magenta without the use of arsenic acid is as follows:—He heats together pure aniline, nitrotoluene, hydrochloric acid, and a small quantity of finely divided metallic iron, to a temperature of about 400° Fahr. (204° C.). for several hours. The pasty mixture soon solidifies to a friable mass resembling crude aniline red.

Dr. HOFMAN and Mr. NICHOLSON have demonstrated that pure aniline, from whatever source obtained, is incapable of furnishing a red dye, but that it does so when mixed with toluidine—toluidine by itself being equally incapable of yielding it.

Magenta consists of brilliant crystals, having a beautiful golden-green metallic lustre, and soluble in water to an intense purplish-red solution. It is a salt of a colourless base, rosaniline, which is prepared from magenta by boiling with hydrate of potassium, and allowing the solution to cool, when it crystallizes out in colourless crystals, having the formulæ $C_{20}H_{19}N_3H_2O$.

Sugar, previously dyed with magenta, is sometimes used as an adulterant of crystallized magenta. The best method of testing magenta is to make a comparative dyeing experiment with the sample under examination, and with one of known purity, using white woollen yarn.

From magenta or hydrochlorate of rosaniline a large number of colouring matters are produced, the most important of which will be briefly described below.

Aldehyde Green.—Prepared by dissolving 1 part of magenta in 3 parts of sulphuric acid, diluted with 1 part of water, adding by degrees $1\frac{1}{2}$ parts of aldehyde, and heating the whole on a water-bath until a drop put in water turns a fine blue. It is then poured into a large quantity of hot water containing 3 parts of hyposulphite of sodium, boiled and filtered. The filtrate contains the green.

Iodine Green.—Produced during the manufacture of the HOFMANN colours; is used for dyeing cotton and silk; its colour being bluer than that of aldehyde green, it is more useful. Iodine green, not being precipitated by carbonate of sodium, is usually sold in alcoholic solution.

Perkins Green.—This is also a magenta derivative, as it is prepared from Britannia violet. It was much used at one time, as it is comparatively a fast colour.

Britannia Violet.—This is obtained in the same manner as the HOFMANN violets, by acting on a solution of magenta in wood spirit, with a thick, viscid fluid of the formula

$C_{10}H_{15}Br_3$, obtained by cautiously acting with bromine on oil of turpentine. It is a beautiful violet, capable of being manufactured of every shade, from purple to blue, and was at one time extensively used.

Hofmann Violets.—On a large scale these violets are produced in deep cast-iron pots, surrounded by a steam jacket, and provided with a lid, having a perforation for distilling over the excess of reagents.

These vessels are charged with a solution of magenta in methylated or wood spirit, and iodide of ethyl or methyl, in proportions according to the shade required, and the whole heated by steam for five or six hours, when the excess of alcohol and iodide of ethyl is distilled over. The resulting product is dissolved in water, filtered, precipitated with common salt, and well washed. They are all moderately fast on wool and silk, although less so on cotton, and, as they can be produced in nearly every shade of violet, are in great use.

Violet Imperial.—If the action of the aniline and magenta in the process of manufacturing aniline blue be stopped before it is finished, and the resulting product treated with dilute acid, a colouring matter called violet imperial is obtained. This dye is now replaced by the HOFMANN violets.

NICHOLSON obtains another violet, Regina purple, from aniline red, by heating it in a suitable apparatus to a temperature between 300° and 327° Fahr. (200° and 215° C.). The resulting mass is exhausted with acetic acid, and the deep violet solution diluted with enough alcohol to give the dye a convenient strength.

The following processes have also been proposed for the production of aniline violet:—

1. WILLIAMS.—Oxidation of an aniline salt by means of a solution of permanganate of potassium.

2. SMITH.—Oxidation of an aniline salt by means of a solution of ferricyanide of potassium.

3. BOLLEY, BEALE, & KIRKMANN.—Oxidation of a cold and dilute solution of hydrochlorate of aniline, by means of a dilute solution of chloride of lime.

4. PRICE.—Oxidation of a salt of aniline by means of peroxide of lead under the influence of an acid.

5. KAY.—Oxidation of a salt of aniline in an aqueous solution of peroxide of manganese.

6. SMITH. - Oxidation of a salt of aniline by free chlorine or free hypochlorous acid.

Benzyl Violet is prepared from magenta, in the same manner as the HOFMANN violet, but the iodide of methyl is replaced by chloride of benzyl, which is obtained by the action of chlorine on toluol.

Spiller's Purple is a shade that is produced by the action of aniline on a red shade HOFMANN violet.

Naphthyl Violet is obtained by heating magenta with naphthylamine.

Aniline Yellow.—Amongst the secondary products obtained during the preparation of aniline red, there occurs a well defined base of a splendid yellow colour, to which the name chrysaniline or phosphine has been given. An inferior quality is also obtained from the residues after the magenta has been extracted from the crude product.

Anthracen, $C_{14}H_{10}$. Anthracen is one of the last products passing over in the dry distillation of coal-tar. Dr. CALVERT says it is " found most abundantly in the 10 or 15 per cent., which comes over between the temperature at which soft pitch is produced, and that at which hard pitch is formed."

Coal-tar contains very variable quantities of anthracen, those tars procured from coals which are richest in naphtha yielding it most abundantly. The coals of South Staffordshire give the largest yield, whilst the Newcastle coals give very little.

GESSERT prepares anthracen from coal-tar as follows :—He places the last pasty portions (the "green grease") of the coal-tar distillation (which must not be carried beyond the point at which soft pitch is formed) first in a centrifugal machine, and then in a hydraulic press at 104° Fahr. (40° C.), or subjects the mass heated to 86° or 104° Fahr. (30° or 40° C.), directly to pressure in a filter-press. The pressed mass consists of about 60 per cent. of anthracen; for further purification it is boiled with light tar-oil or petroleum naphtha, and finally heated till it melts. The residue contains 95 per cent. of anthracen.

The following method for the purification of crude anthracen, contaminated with oily matters, is by SCHULLER: The crude anthracen is carefully heated to commencing ebullition in a capacious retort connected with a tubulated receiver of glass or earthenware, the lower aperture of which is closed with a fine wire sieve. A strong current of air is then blown into the retort with a pair of bellows, whereby the anthracen is driven over in a very short time nearly pure and dry, and condenses in the receiver as a faintly yellowish, snowy mass. By this method a quantity of anthracen, the purification of which by re-crystallization or sublimation would take several days, may be purified in as many hours; moreover it is obtained in a pulverulent form, in which it is very readily acted on by oxidizing agents. Anthraquinon, prepared from crude anthracen, may also be obtained by this method in the form of a light yellow powder, resembling flowers of sulphur.

If it is required to obtain anthracen very pure (or for treatment by the di-chlor process, in which case it is essential that it be free from the higher bodies), it must, before being placed in the retorts, be ground up with 12 to 20 per cent.

of Montreal potashes and a small quantity of lime, as originally recommended by PERKIN. This operation causes no loss of anthracen, as it has been proved that caustic potash is without action on it, but other substances, which no amount of washing will remove, are entirely split up by the alkali. Caustic soda is quite useless for the purpose. Anthracen thus treated yields, when crystallized from benzol, the pure substance in the form of fluorescent transparent crystals, consisting of four or six-sided plates, which, when seen by transmitted light, are of a very pale blue colour, but of a pale violet by reflected light.

A method for determining the amount of pure anthracen either in commercial anthracen or in crude green grease is the following :—The melting-point of the sample in question is first determined; 5 to 10 grammes are sufficient for the operation. It is put between thick folds of blotting paper, and placed under a press, between plates which have been previously warmed. The anthracen remaining upon the paper after pressure is weighed. The residue after it has been boiled with a certain quantity of alcohol, filtered, washed with cold alcohol and dried, is weighed as pure anthracen. It is now advisable to determine the melting-point of the purified product, which will generally be $318°$ Fahr. ($210°$ C.).

The test which is now almost universally adopted is LUCK's test, which consists in boiling a gramme of the sample, dissolved in glacial acetic acid, with an excess of solution of chromic acid. When the solution is allowed to cool and mixed with water, the anthraquinon crystallizes out, this is washed and dried, and the subsequent treatment which is adopted consists in dissolving the anthraquinon in 10 grammes of warm concentrated sulphuric acid, and then (after thorough cooling) separating out the purified product by the addition of water. From the weight of this, after washing and drying, the quantity of the original sample of anthracen is calculated. All samples should be washed with cold petroleum spirit, and pressed and dried before the gramme is weighed out for testing, as by that means spurious samples are at once detected, and genuine samples are obtained in the form of an even and compact mass. Anthracen is only slightly soluble in alcohol, but rather more so in ether and bisulphide of carbon. It is more soluble in hot, but less so in cold benzene. Petroleum boiling between 160° and 195° Fahr. dissolves less than benzene.

"Anthracene dissolves in concentrated sulphuric acid with a green colour, and forms conjugated monsulpho- or bisulpho-anthracene acid, according to the temperature employed. Chlorine and bromine give rise to substitution products. Nitric acid acts on it with great violence, with formation of anthraquinone, nitro-anthraquinone, and other compounds, according to the temperature and proportion of the substances taken. With picric acid anthracene forms a compound, crystallizing in very bright ruby-red needles, which, by the aid of the microscope, are seen to be prisms. To prepare it a saturated solution of picric acid in water at 80° Fahr. (26·6° C.), is mixed with a saturated solution of anthracene in boiling alcohol; on cooling the compound is deposited in the crystalline state. It is rapidly decomposed, by an excess of alcohol, into picric acid and anthracene, the solution assuming a yellow tint. This reaction can be employed to distinguish anthracene from naphthalene and other hydrocarbons, naphthalene under similar circumstances forming a compound which crystallizes in fine golden-yellow needles, whilst chrysene gives rise to clusters of very small yellow needles."[*] Another characteristic of anthracen,

---

[*] CALVERT's "Dyeing and Calico-Printing," edited by STENHOUSE and GROVES.

noticed by FRITZSCHE, is its deportment under the microscope with a solution of binitro-anthraquinon in benzene. In this reaction fine rhomboidal scales of a beautiful pink colour are formed, the purity and brilliancy of the colour depending on the purity of the anthracen.

In the "Bul. Soc. Chim.," vii. 274, several reactions by which anthracen is formed are described by BERTHELOT, such as by the action of heat on other hydrocarbons, or by passing the vapours of ethylene, styrolene, and benzene through a porcelain tube heated to bright redness.

A great number of products are procured from anthracen, by far the most important of these being artificial alizarin.

Alizarin, Artificial, $C_{14}H_8O_4$. This colour was first obtained by GRAEBE and LIEBERMANN in 1869 from anthraquinon, an oxidation product of anthracen, this latter being, as already stated, a substance which is formed during the destructive distillation of coal-tar. These chemists converted anthracen into anthraquinon by means of nitric acid.

According to the original method of preparing alizarin, the anthraquinon was first converted into a dibromide of anthraquinon by treatment with bromine, and this bromated compound, by further treatment either with caustic potash or soda at a temperature of 356° to 392° Fahr. (180° to 200° Cent.) converted into alizarin-potassium (or alizarin-sodium if caustic soda has been used), from which the alizarin is set free by means of hydrochloric acid.

Alizarin is now procured from anthraquinon by treatment at a temperature of 500° Fahr. (260 Cent.), with concentrated sulphuric acid of 1·84 sp. gr., the anthraquinon being converted into a sulpho-acid; this acid is next neutralized with carbonate of lime, the fluid decanted from the deposited sulphate of lime, and carbonate of soda added to it, with the object of throwing down all the lime. The clear liquid is then evaporated to dryness, the resulting saline mass is converted into alizarin-sodium by heating it with caustic sodium. From the alizarin-sodium thus obtained the alizarin is set free by the aid of hydrochloric acid.

In another method the preparation of anthraquinon is avoided, and anthracen employed directly, by first converting it, by means of sulphuric acid and heat, into anthracensulphonic acid. After having been diluted with water, the solution of this acid is treated with oxidizing agents (peroxides of manganese and lead, chromic acid, nitric acid), and the acid fluid is afterwards neutralized with carbonate of lime. When peroxide of manganese has been used, the manganese is precipitated as oxide. The oxidized sulpho-acid having been previously converted into a potassium salt, the latter being heated with caustic soda, alizarin is obtained. The details of these two processes will be found set forth in the terms of the patent taken out by Messrs. CARO, GRAEBE, & LIEBERMANN, further on.

The following method of preparing alizarin from anthracen is by GIRARD. The material used is that which distils between 290° and 360° Cent.; it is purified by distillation and pressure, the portion which passes over, between 300° and 305° Cent., being collected separately. This mixture is treated with potassium chlorate and hydrochloric acid, whereby it is converted into tetra-chlorinated products. These are oxidized either by nitric acid in the water-bath, or by a metallic oxide (red or brown oxide of lead,) and sulphuric or acetic acid. In the first place a mixture of dichloranthraquinon and chloride of chloroxyanthranyl are obtained. These substances are treated in presence of a metallic oxide (oxide of zinc, oxide of copper, or litharge) with an alcoholic solution of sodium acetate. The metallic oxide removes the last atom of chlorine from the sodium chloroxyanthranilate, and converts it, like the dichloranthraquinon, into alizarin. The purification is effected by means of benzene, petroleum,

&c., which dissolve out the foreign matters, and by successive precipitation from the alkaline solutions by mineral acids. The foreign matters may also be separated by means of a little alum, when it is necessary to work with neutral potash or soda salts.

Another method for the preparation of alizarin has been patented by DALE & SCHORLEMMER. It is as follows :—1 part of anthracen is boiled with from 4 to 10 parts of strong sulphuric acid, then diluted with water, and the solution neutralized with carbonate of calcium, barium, potassium, or sodium. The resulting sulphates having been removed by filtration or crystallization, the solution is heated to between 180° and 260° Cent. with caustic potash or soda, to which a quantity of potassium nitrate or chlorate has been added, about equal in weight to the anthracen, as long as a blue-violet colour is thereby produced. From this product the alizarin is separated in the usual way by precipitation with an acid. Several other patents have been taken out for the preparation of artificial alizarin.

The specification of Messrs. CARO, GRAEBE, & LIEBERMANN, and dated June 25, 1869, was the first which was taken out in England. We quote it here because it enters more fully into detail than any of the others.

"Our invention is carried into effect by means of either of the two processes which we will proceed to describe.

"In the one process we proceed as follows :—We take about 1 part by weight of anthraquinone and about 3 parts by weight of sulphuric acid of about specific gravity 1·848, and introduce the same into a retort, and the contents are then to be heated up to about 260° Cent., and the temperature is maintained until the mixture is found no longer to contain any appreciable quantity of unaltered anthraquinone. The completion of this operation may be ascertained or tested by withdrawing a small portion of the product from time to time, and continuing the operation at the high temperature until such product, upon being diluted with water, is found to form a substantially perfect solution, thereby indicating that the anthraquinone has become either entirely or in greater part converted into the desired product. The products thus obtained are then allowed to cool, and are diluted with water; carbonate of lime is then added in order to neutralize and remove the excess of sulphuric acid contained in the solution; the mixture is then filtered, and to the filtrate, carbonate of soda, by preference in solution, is to be added until carbonate of lime is no longer precipitated; the mixture is then filtered, and the clear solution is evaporated to dryness, by which means the soda salt of the sulpho-acids of anthraquinone are obtained, and which are to be treated in the following manner:—We take about 1 part by weight of this product, and from 2 to 3 parts by weight of solid caustic soda, water may be added or not, but by preference we add as much water as is necessary to dissolve the alkali after admixture; we heat the whole in a suitable vessel, and the heating operation is continued at a temperature of from about 180° to 260° Cent., for about one hour, or until a portion of the mixture is found upon withdrawing and testing it, to give a solution in water, which being acidulated with an acid—for example, sulphuric acid—will give a copious precipitate of the colouring matters. The heating operation having been found to have been continued for a sufficient time, the resulting products are then dissolved in water, and we either filter or decant the solution of the same, from which we precipitate the colouring matters or artificial alizarine, by means of an acid, such, for example, as sulphuric or acetic acid. The precipitated colouring matters thus obtained are collected in a filter or otherwise, and after having been washed may be employed for the purpose of dyeing and printing, either in the same way as preparations of madder are now used or otherwise.

"In carrying out our other process, we proceed as follows: —We take about 1 part by weight of anthracene, and about 4 parts by weight of sulphuric acid of specific gravity of about 1·848, and the mixture being contained in a suitable vessel, is heated to a temperature of about 100° Cent., and which temperature is to be maintained for the space of about three hours; the temperature is then to be raised to about 150° Cent., which temperature is to be maintained for about one hour, or until a small portion of the product when submitted to the two subsequent processes hereinafter described, is found to produce the desired colouring matters; we then allow the result obtained by this operation to cool, and dilute it with water, by preference in the proportion of about three times its weight. To the solution thus obtained we add for every part of anthracene by weight which had been employed in the previous operations, from about 2 to 3 parts by weight of peroxide of manganese, preferring to employ an excess, and we boil the whole strongly for some time, and in order fully to ensure the desired degree of oxidation, the mixture may be subsequently concentrated, and by preference be evaporated to dryness, and the heat be continued until a small portion of the oxidized product, when submitted to the subsequent processes hereinafter described, will produce the desired colouring matters. We then neutralize and remove the sulphuric acid contained in this mixture, and at the same time precipitate any oxides of manganese that may be held in solution, by adding an excess of caustic lime, which we use by preference in the form of milk of lime, and we add the same until the mixture has an alkaline reaction. We then filter, and add to the filtrate carbonate of soda, until there is no further precipitation of carbonate of lime. The solution is then filtered and evaporated to dryness, and we thus obtain the potash or soda salts of what we call the sulpho-acids of anthraquinone.

"In effecting the conversion of the oxidized products thus obtained into colouring matters, or into what we call artificial alizarine, we proceed as follows:—We take 1 part by weight of this product, and from 2 to 3 parts by weight of solid caustic soda, and water may be added or not, but by preference we add as much water as may be necessary to dissolve the alkali. After admixture we heat the whole in a suitable vessel, and continue the heating operation at a temperature of about 180° to about 260° Cent. for about one hour, or until a portion of the mixture is found to give a solution in water, which upon acidulation with an acid, for example, sulphuric acid, is found to give a copious precipitate of the colouring matters. The heating operation having been found to have been continued for a sufficient time, we then dissolve the product in water, and either filter or decant the solution of the same, from which we precipitate the colouring matters or artificial alizarine by means of a mineral or organic acid, such, for example, as sulphuric or acetic acid.

"Instead of acting upon anthracene, by means of sulphuric acid of the density before mentioned, fuming sulphuric acid may be employed, but we prefer to use the ordinary kind before described.*

"In order to effect the process of oxidation, before referred to, other oxidizing agents may be used in the place of the oxide of manganese, before mentioned, such, for example, as peroxide of lead, or chromic, nitric, or other acids capable of effecting the desired oxidation may be employed."

---

* The preference here mentioned is for reasons of economy, but now that solid sulphuric acid is manufactured by Messrs. MESSEL, CHAPMAN & Co., it is used in the place of ordinary sulphuric acid.

Mr. W. H. PERKIN's patent is similar in principle to that of Messrs. CARO, GRAEBE, & LIEBERMANN, and is dated only one day later.

The following is an outline of a patent taken out in France in May, 1869, by MM. BRŒNNER & GUTZKON for the manufacture of artificial alizarin. One part of anthracen is heated with 2 parts of nitric acid, sp. gr. 1·3 to 1·5. The anthraquinon thus produced is washed and dissolved at a moderate heat in sulphuric acid. Mercuric nitrate is now added, which converts the anthraquinon into alizarin. The mass thus formed is dissolved in an excess of alkali, which precipitates the oxide of mercury, and retains the colouring matters in solution. The alkaline liquor is decanted and neutralized with sulphuric acid, and the precipitate thus formed is washed and collected. If not quite pure, the treatment with alkali must be repeated. The complete specification of this patent is published in the *Moniteur Scientifique*, vol. xi. p. 865.

Bromine, by its action on alizarin, produces a derivative which gives rather redder shades, but if nitrous acid vapours act on dry alizarin they produce nitro-alizarin, which is known commercially as alizarin-orange.

Alizarin-blue is obtained by heating a mixture of alizarin-orange, glycerin and sulphuric acid.

In England a large quantity of artificial alizarin is manufactured by the process of Mr. PERKIN, by Messrs. BURT, BOULTON, & HAYWOOD, and is used as a substitute for madder and madder extract, in Turkey-red dyeing and topical styles. The largest makers of artificial alizarin on the Continent are Messrs. GESSERT FRÈRES, of Elberfeld, Messrs. MEISTER, LUCIUS & Co., of Höchst, near Frankfort, and the BADISCHE ANILIN UND SODA FABRIC, Mannheim.

Anthrapurpurin, $C_{14}H_6O_6$. — A colouring matter obtained as a secondary product in the preparation of alizarin from anthracen. It may be prepared by dissolving the crude colouring matter in a dilute solution of carbonate of soda, and shaking up the resulting solution with freshly precipitated alumina, which combines with the alizarin, leaving the anthrapurpurin in solution. This is filtered off from the alizarin lake, heated to boiling, and acidified with hydrochloric acid. The colouring matter which is precipitated is thrown on to a filter, washed and dried.

Anthrapurpurin is the principal product of PERKIN's di-chlor process, of which the following is a brief sketch:— The anthracen, which had been purified by washing and distillation with potash, was placed in charges of 400 lbs. in leaden ovens, which rested on shallow wrought-iron steam-chests (set in pairs), and chlorine was passed over. The crude di-chlor-anthracen thus obtained was then drawn out into shallow open tubs and stirred up with cold heavy naphtha. After being allowed to stand for some time it was pressed, then again washed with clean naphtha and again pressed. After that it was placed in shallow trays and dried at a moderate heat. The subsequent operations for converting the di-chlor-anthracen into red-shade alizarin are the same as those adopted in the treatment of anthraquinon.

Anthrapurpurin has about the same affinity for mordants as alizarin. It forms red with alumina, and purple and black with iron mordants. The reds are much purer and less blue in colour than those of the alizarin, whilst the purples are not so good and the blacks more intense. The anthrapurpurin colours resist soap and light quite as effectively as those produced with alizarin. When employed to dye Turkey-red, anthrapurpurin gives a very brilliant scarlet shade of colour, which is of remarkable durability.

Besides the products obtained from aniline, a series

of colours have been obtained from phenol, or carbolic acid, another substance obtained from coal-tar.

Picric Acid.— This is obtained by treating in a suitable apparatus, with proper precautions, carbolic acid with nitric acid. It is a pale yellow crystalline acid, forming dark orange explosive salts, and dyeing silk a fine yellow.

Isopurpurate of Potassium. *Syn.* PICRIC RED.—By treating picric acid with cyanide of potassium a very explosive salt is obtained, used to dye wool a deep red colour.

Grenate Brown is the name given to a colouring matter which is essentially a crude isopurpurate. When dry, grenate brown explodes if subjected to the smallest amount of friction. It should therefore be kept in the form of a paste, which may be preserved moist by means of glycerin.

Phœnicienne is a deep garnet colour, obtained by the action of nitro-sulphuric acid on carbolic acid. It furnishes a variety of solid shades which resist sunlight, and even chloride of lime; they surpass in purity and brightness all similar colours obtained with dye-woods and orchil.

Aurine. *Syn.* ROSOLIC ACID.— This is obtained by heating a mixture of sulphuric, oxalic, and carbolic acids, and purifying the product. It is a beautiful reddish, resinous substance, with a pale green lustre, and yielding an orange-coloured solution, changed by alkalies to a splendid crimson. Owing to the difficulty in using it, however, it is not very extensively employed.

Peonine. *Syn.* CORALLINE.—This dye is formed when rosolic acid and ammonia are heated to between 248° and 284° Fahr. (120° to 140° Cent.). It is a fine crimson dye, forming shades similar to safflower, on silk, but, owing to the bad effects of acids, not much used.

Azuline.— Prepared by heating coralline and aniline together. A coppery coloured resinous substance, soluble in alcohol, and with difficulty in water, and dyeing silk a blue colour. The aniline blues, however, have superseded it to a great extent.

Carbolic acid, when saturated with ammonia and heated, yields aniline, and DUSART & BARDY have found that when heated under pressure with sal-ammoniac and hydrochloric acid, the principal yield is diphenylamine. This substance by treatment with oxalic acid (there are many other substances which can be used, but they are not so advantageous), yields a splendid blue called Diphenylamine Blue.

Naphthalene occurs in such abundance in coal-tar that the supply is far in excess of the demand. By treating this in exactly the same manner as benzol is converted into aniline, a solid crystalline white base, termed naphthylamine, is produced. From this substance are obtained the following dyes:—

DINITRONAPHTHOL. *Syn*. MANCHESTER YELLOW, is prepared direct without separating the intermediate product (naphthol) by acting on the result of the action of nitrite of soda or hydrochlorate of naphthylamine with nitric acid at a boiling heat. Ammonia is added to the liquid when cold, and the compound thus produced is purified by crystallization. In steam dyeing, Manchester yellow has this advantage over picric acid—the latter is volatilized by the heat, whilst the former is fixed to the fabric. If nitro-naphthalene is mixed with slaked lime and a solution of caustic potash, and the mass is heated for twelve hours to 300° F. (149° Cent.) in a current of air, the colouring matter, which is separated by adding acid to an aqueous extract of the product, is a yellow dye termed French Yellow.

Naphthalene treated with a mixture of chlorate of potash and hydrochloric acid yields two substances, one of which, when treated with nitric acid, produces phthalic acid, which by the action of heat is converted into phthalic anhydride. This substance when treated with resorcine (formerly prepared from Brazil wood, but now made from benzol) yields a splendid yellow dye, named Fluorescein. This, if ground in alcohol and carefully treated with bromine, yields the brilliant scarlet dye named Yellowish Eosin.

A naphthalene red, known as MAGDALA RED, was obtained by SCHIENDL, in 1867. It possesses a tinctorial value equal to fuchsin, which latter colour it excels in the property of fastness. It is procured by first acting on naphthylamine with nitrous acid, and then treating the resulting product with naphthylamine, which gives rise to Magdala Red.

SAFROSIN is obtained by the action of nitrate of soda and sulphuric acid on a solution of yellowish eosin in boiling water. This colour gives a more bluish-red tinge than eosin. The other substance produced in the manufacture of phthalic acid is an acid which, combined with a base, yields an orange dye.

APPENDIX.

Table of the principal Weights and Measures of the Metrical System, with their Equivalents in Common Weights and Measures.

LENGTH.

Metrical.		In English inches.	In English feet = 12 inches.	In English yards = 3 feet.
Millimetre	=	0·03937	0·0032809	0·0010936
Centimetre	=	0·39371	0·0328090	0·0109363
Decimetre	=	3·93708	0·3280899	0·1093633
Metre	=	39·37079	3·2808992	1·0936331
Decametre	=	393·70790	32·8089920	10·9303310
Hectometre	=	3937·07900	328·0899200	109·3633100
Kilometre	=	39370·7900	3280·899200	1093·6331000

CAPACITY.

Metrical.		In cubic inches.	In cubic feet = 1728 cubic inches.	In pints = 34·65923 cubic inches.	In gallons = 8 pints = 277·27384 cubic inches.
Millilitre, or cubic centimetre (c.c.)	=	0·061027	0·0000353	0·001761	0·00022010
Centilitre, or 10 cubic centimetres	=	0·610271	0·0003532	0·017608	0·00220097
Decilitre, or 100 cubic centimetres	=	6·102705	0·0035317	0·176077	0·02200967
Litre, or cubic decimetre	=	61·027052	0·0353166	1·760773	0·22009668
Decalitre, or centistere	=	610·270515	0·3531658	17·607734	2·20096677
Hectolitre, or decistere	=	6102·705152	3·5316581	176·077341	22·0096767

WEIGHT.

Metrical.		In English grains.	In Troy ounces = 480 grains.	In Avoirdupois lbs. = 7,000 grains.	In cwts. = 112 lbs. = 734,000 grains.
Milligramme	=	0·015432	0·000032	0·0000022	0·00000002
Centigramme	=	0·154323	0·000322	0·0000220	0·00000020
Decigramme	=	1·543235	0·003215	0·0002205	0·00000197
Gramme	=	15·432349	0·032151	0·0022046	0·00001968
Decagramme	=	154·323488	0·321507	0·0220462	0·00019684
Hectogramme	=	1543·234880	3·215073	0·2204621	0·00196841
Kilogramme (kilo)	=	15432·348800	32·150727	2·2046213	0·01968412

Specific Gravities corresponding to Degrees of
BAUMÉ's *Hydrometer.*
For Liquids heavier than Water (POGGIALE).

Degrees.	Specific Gravity.	Degrees.	Specific Gravity.	Degrees.	Specific Gravity.	Degrees.	Specific Gravity.
0	1·000	20	1·161	40	1·383	60	1·711
1	1·007	21	1·171	41	1·397	61	1·732
2	1·014	22	1·180	42	1·410	62	1·753
3	1·022	23	1·190	43	1·424	63	1·774
4	1·029	24	1·199	44	1·438	64	1·796
5	1·036	25	1·210	45	1·453	65	1·819
6	1·044	26	1·221	46	1·468	66	1·846
7	1·052	27	1·231	47	1·483	67	1·872
8	1·060	28	1·242	48	1·498	68	1·897
9	1·067	29	1·253	49	1·514	69	1·921
10	1·075	30	1·264	50	1·530	70	1·946
11	1·083	31	1·275	51	1·546	71	1·974
12	1·091	32	1·286	52	1·563	72	2·000
13	1·100	33	1·297	53	1·580	73	2·031
14	1·108	34	1·309	54	1·597	74	2·059
15	1·116	35	1·320	55	1·615		
16	1·125	36	1·332	56	1·634		
17	1·134	37	1·345	57	1·652		
18	1·143	38	1·357	58	1·671		
19	1·152	39	1·370	59	1·691		

Specific Gravities corresponding to Degrees of
BAUMÉ's *Hydrometer.*
For Liquids lighter than Water (FRANCŒUR).

Degrees.	Specific Gravity.	Degrees.	Specific Gravity.	Degrees.	Specific Gravity.	Degrees.	Specific Gravity.
10	1·000	23	0·918	36	0·849	49	0·789
11	0·993	24	0·913	37	0·844	50	0·785
12	0·986	25	0·907	38	0·839	51	0·781
13	0·980	26	0·901	39	0·834	52	0·777
14	0·973	27	0·896	40	0·830	53	0·773
15	0·967	28	0·890	41	0·825	54	0·768
16	0·960	29	0·885	42	0·820	55	0·764
17	0·954	30	0·880	43	0·816	56	0·760
18	0·948	31	0·874	44	0·811	57	0·757
19	0·942	32	0·869	45	0·807	58	0·753
20	0·936	33	0·864	46	0·802	59	0·749
21	0·930	34	0·859	47	0·798	60	0·745
22	0·924	35	0·854	48	0·794		

The temperature at which these instruments were originally adjusted by BAUMÉ was 12·5° Cent. (54·5° Fahr.). They are now commonly adjusted in this country at 58° or 60° Fahr.

The degrees of TWADDLE's hydrometer may be converted into the corresponding specific gravities by multiplying them by 5 and adding 1000.

INDEX.

A.

Acid, 102
Acid discharge 102
Adjective colours, 36
Ageing, 71
    machine, 71
Aldehyde green, 175
Alizarin, 153
    artificial, 181
Alizarin, artificial, Brœnner & Gutzkon's patent for preparing, 187
Alizarin, artificial, Caro, Graebe & Liebermann's method of preparing, 183
Alizarin, artificial, Dale & Schorlemmer's method of preparing, 183
Alizarin, artificial, preparation of, 181, 182
Alizarin, artificial, by Girard's method, 182
Alizarin, artificial, Perkin's method of preparing, 187
Alizarin blue, 187
Alkermes, 147
Aloes, 133
Aloin, 133
Alumina, nitrate of, for steam styles, 113
Amber for steam styles, 121
Ammoniacal cochineal, 60
Ammonia purpurate, 156
Aniline, 161

Aniline, from coal-tar, 161
    from nitro-benzol (Zinin), 162
    from nitro-benzol (Béchamps), 162, 163
    from indigo, 164
    other methods of preparing, 164, 165
    properties of, 165
    tests for, 165
    Reimann's test for, 166
    black, 167
    black with vanadium, 45
    blue, 168
    blue dye, 49
    blue for printing, 170
    brown, 170
    green, 171
    grey, 170
    pink, 170
    red, 172
Aniline red, Coupier's method of preparing, 174
Aniline red, Dale & Caro's patent, 174
Aniline red, Gerber-Keller's patent for preparation of, 172
Aniline red, Laurent & Casthélaz's method of preparing, 174
Aniline red, Renard's method of preparing, 174
Aniline red, Lauth & Depouilly's method for preparing, 172
Aniline red, Medlock's patent for preparation of, 172

INDEX.

Aniline red, Smith's method of preparation, 174
  violets, 176, 177
Aniline violet, by Smith's method, 177
Aniline violet, by Bolley & Beale, 177
Aniline violet, by Price's method, 177
Aniline violet, by Kay's method, 177
Aniline violet and blue dyes, Girard & De Laire, 174
Aniline, yellow, 177
Annotta, 133
  adulteration of, 135
  to test, 134
Anthracen, 177
  Luck's test for, 179
  properties of, 180
  Schuller's method of purifying, 178
Anthrapurpurin, 187
Antichlore, 29
Apparatus, dyeing, 39
Archil, 135
Aurine, 189
Avignon berries, 157
Azaleine, 172
Azuline, 189

B.

Bark, Quercitron, 158
  standard for steam styles, 117
Barwood, 136
  red, for cotton, 53
Basic tin compound for steam styles, 117
Bastard saffron, 158
Benzol, 160
Benzyl violet, 177
Bergman's theory, 38
Berries, 157
  Avignon, 157
  French, 157
  Persian, 157
  yellow, 157
Bismark's brown, 170
Bixin, 134

Black aniline (calico printing) 96
  aniline with vanadium, 45
  common, 44, 45
  chrome, 45
  De Vinant's, 46
  dyes for cotton, 43, 44, 45, 46, 47
  for extract styles, 129
  Italian, 45
  liquor, 155
  for machine work, 95
  for pigment colours, 126
  for spirit styles, 122
  standard for steam styles, 118
  dyes for wool, 55, 56
Blankets for cylinder printing, 67
Bleaching, chemico-mechanical process of, 21, 22, 23, 24
  of cotton, 8
  feathers, 28
  history of, 1, 2, 3, 4, 5, 6, 7
  Hodge's process of, 21, 22, 23, 24
  linen, 13, 14, 15, 16
  materials for paper, 28, 29
  new or continuous process of, 9, 10, 11, 12
  powder, proposed substitutes for, 12, 13, 32
  printed paper, 29
  Ramsay's method, 32
  silk, 27
  straw, 30
  Tessié du Mathey's method, 32
  theory of, 31
  wax, 30, 31
  woollen goods, 26, 27
  yarn, 21
Block printing, 63
Blue, alizarin, 187
  black for wool, 56
  cerulean, for extract styles, 131
  dyes for cotton, 49
  diphenylamine, 190
  discharge on bronze, 124
  for indigo styles, 89, 90
  for spirit styles, 122
  dark, for steam styles, 114, 121
  medium, for steam styles, 115

INDEX.

Blue, pale, for steam styles, 115, 121
   (methyline) for extract styles, 131
   fast, standard, 104
   fast, for block work, 104
   de Lyons, 168
   opal, 168
   de Paris, 171
   standard, for pigment colours, 127
   dyes for silk, 55
   standard, for steam styles, 118, 119
Bolley & Beale's method of preparing aniline violet, 177
Bousage, 72
Brazil wood, 138
Brazilin, 139
Brightening printed goods, 75
Britannia violet, 175
Brœnner & Gutzkon's method of preparing artificial alizarin, 187
Bronze colour style, 93
Brown, Bismark, 47
   bright, 48
   catechu, 47
   dark, 48
   dyes for cotton, 47, 48
   grenate, 189
   for indigo styles, 110
   light, 48
   light, for indigo styles, 111
   madder, to resist heavy covers of purple, 98
   standard, calico printing, 97
   for machine work, 97
   medium, 48
   medium (calico printing) 98
   for pigment colours, 125
   medium, for steam styles, 117
   dye for wool, 57
Buff, for indigo styles, 110
   for pigment styles, 125
   (medium), for steam styles, 119
   (pale) for steam styles, 119
   standard, 103
   standard, for steam styles, 119

C.

Calico Printing, 61
   history of, 61, 62
   styles of, 86
Campeachy wood, 149
Camwood, 138
Canary dye for cotton, 51
Caro, Graebe & Liebermann's method of preparing artificial alizarin, 183
Carthamin, 158
Catechu, 137
   varieties of, 137
   to test, 137
   brown, 47
Chamois dye for cotton, 50
   for indigo, 110
   for steam styles, 118
Chemico-mechanical process of bleaching, 21, 22, 23, 24
Chestnut for steam styles, 118
Chevreul's theory, 38
China blue style, 90
Chinoline blue, 139
Chlor-machine, 106
Chlorine, Dobbie and Hutcheson's process for obtaining, 20
Chocolate for machine work, 98
   (for steam styles), 114, 121
   dye for wool, 57
Chrome, arseniate of, standard, 104
   black, 45
   chloride of, standard, 103
   nitro-acetate of, for extract styles, 129
   sulphate of, standard, 103
   yellow for cotton, 54
Chryso-rhamnin, 157
Cinchonine, 139
Cinnamon for steam styles, 118
Claret dye for cotton, 51
   dye for wool, 57
Cleansing process, 74
Coal-tar, 160
   colours, 160
   colours, how to dye cotton with, 52
Cochineal, 139
   varieties of, 140
   adulteration of, 140

Cochineal, to test, 140
  ammoniacal, 60
  ammoniacal, standard for steam styles, 121
Colours, adjective, 36
  substantive, 36
Colour doctors, 67
Colouring matters and mordants, nature of union between, 38
Coralline, 189
Cotton and calico cleansing, 69
Cotton, bleaching of, 8
  forms in which dyed, 35
  impurities in, 9
  and calico, singeing of, 68
  black dyes for, 43, 44, 45, 46, 47
  blue dyes for, 49
  brown dyes for, 47, 48
  canary dye for, 51
  chamois dye for, 50
  claret dye for, 51
  and coal-tar colours, 52
  drab dye for, 51
  green dye for, 49
  dark mauve dye for, 50
  orange dye for, 54
  pink dyes for, 53
  red dye for, 53
  slate dye for, 50
  yellow dye for, 54
  violet dye for, 50
Covers, 101
Crum's theory, 38
Cudbear, 140
Cuir, for steam styles, 118
Cylinder printing, 64, 65, 66, 67, 68
  apparatus, 64, 65, 66
Cyanine, 139

D.

DALE & SCHORLEMMER's method of preparing alizarin, 183
Dale & Caro's patent for preparing aniline red, 174
Dahlia, 170
Dégommage, 74
De Vinant's black, 46
Di-chlor process, Perkins, 188

Dinitronaphthol, 190
Dip, the, 20
Diphenylamine blue, 106
Discharge style, 90
  white for indigoes, 109
Discharges, 101
  various, 91
Dobbie & Hutcheson's process for obtaining chlorine, 208
Drab (calico-printing), 105
  dye for cotton, 51, 52
  bright for cotton, 52
  for machine work, 98
  dye for wool, 57
  for steam styles, 120
Dunging process, 72, 73
Dung substitutes, 72, 73
Dyeing, 33
  apparatus, 39, 40
  history of, 33, 34
  upon mordant style, 87
  printed goods, 74
Dye-becks, 74
Dye stuffs, 133
Dyers' saffron, 158
  woad, 159

E

EMERALDINE, 171
Eosin, yellowish, 191
Extract, or topical fast style, 93
Extracts, madder, 152

F.

FARINA GUM WATER, 99
Fawn (calico-printing), 99, 105
  for steam styles, 118
Feather bleaching, 28
Ferrocyanide of tar for steam styles, 115
Finishing, 76
Five-fourths mordant, 154
Flax, impurities in, 171
Fleur de garance, 151
Flowers of madder, 151
Fluorescine, 191
French berries, 157

INDEX.

French berries, pink style, 88
   yellow, 190
Fuchsine, 172
Fustic, 141
   old, 141
   young, 141
Fustine, 141

G.

GALL LIQUOR, 102
Garance, fleur de, 151
Garanceux, 152
Garancine, 152
   style, 89, 105
Gerber-Keller's patent, for preparation of aniline red, 172
Gessert's method for obtaining anthracen, 178
Girard's method for preparing artificial alizarin, 182
Glycerin standard for extract styles, 130
Green aldehyde, 175
   dye for cotton, 49
   discharge on bronze, 124
   fast (calico printing), 104
   for indigo styles, 110
   iodine, 175
   (dark) for steam styles, 115
   (medium) for steam styles, 121
   (methyl) for extract styles, 131
   (pale) for steam styles, 116
   Perkin's, 175
   for pigment colours, 125
   dyes for wool, 58
Grenate brown, 189
Grey for pigment colours, 126
Gum, how to test, 84
   lead, 105

H.

HABANA BROWN, 170
Hæmatoxylin, 149
Hellot's theory, 38
History of bleaching, 1, 2, 3, 4, 5, 6, 7
   of calico-printing, 61
   of dyeing, 33, 34

Hodge's process of bleaching, 21, 22, 23, 24
Hoffman violets, 176
Hypochlorite of magnesia, 21
   of soda (chloride of soda), 20

I.

INDIAN COCHINEAL, 148
   yellow, 157
Indican, 142
Indiglucin, 143
Indigo, 141, 142
   how prepared for dyeing, 143
   how to test, 143
   discharge style, 90, 107
   blue style, 89
   sulphate of, 146
   soluble, 146
   cold vat, 144
   vat, German, 145
   potash vat, 144
   vat, Schützenberger & De Lalande, 144
Iodine green, 175
Irish plan of bleaching linen, 14, 15, 16
   of bleaching yarns, 24, 25, 26
Iron liquor, 155
Isopurpurate of potassium, 189
Italian black, 45

J.

JET BLACK FOR WOOL, 56

K.

KAY'S METHOD of obtaining aniline violet, 177
Kermes, 147
Kieserite, 21

L.

LAC, 148
   lake, 148
   stick, 148
   tests for, 148, 149

INDEX.

La Kao, 149
Lake, 37
Laurent & Casthélaz's method of preparing aniline red, 174
Lauth & Depouilly's patent for aniline red, 172
Lavender for steam styles, 120
Lead gum, 105
Lepidine, 139
    blue, 139
Le Pileur d'Appligny's theory, 38
Lilac for spirit styles, 122
Lima wood, 139
Lime-juice mixture, 102
Linen, bleaching of, 13, 14, 15
    bleaching, Irish plan, 14, 15, 16
    bleaching, Scotch plan, 14, 15, 16
    cleansing process, 69
Liquor, black, 155
    iron, 155
    red, 153
Logwood, 149
Luck's test for anthracen, 179
Luteolin, 159
Lyons' blue, 168

M.

MACLURIN, 141
Madder, 150
    adulteration of, 155
    extracts, 152
    flowers of, 151
    to test, 155
    varieties of, 150, 151
    and alizarin dyed style, 87
    and alizarin dyed style, blocked, 88
    brown, to resist heavy covers of purple, 98
    purple, dark (calico printing), 96
    purple light (calico printing), 97
    purple medium (calico printing), 97
Magdala red, 191
Magenta, 172
    adulterations of, 155
Magnesia, hypochlorite of, 21

Manchester yellow, 190
Manganese bronze style, 93
Materials for paper, bleaching, 28, 29
Mauve, 169
    or violet discharge on bronze, 124
    dye for cotton, 50
    for extract styles, 130
Mauveine, 169
Medium brown, 48
Medlock's patent for aniline red, 172
Mills, wash, 17
Mordants, 36, 77, 101
    list of, 77
    remarks on, 78, 79, 80, 81
Mordant, alkaline red, 101
    alkaline red, light, 101
    alkaline pink, 101
    alkaline pink, light, 101
    pans, 85
Morine, 141
Moritannic acid, 141
Murexid, 156

N.

NAVY BLUE DYE, 49
Naphthalene, 190
Naphthyl violet, 177
Nicaragua wood, 139
Nicholson's blue, 168
Nitrate of alumina for steam styles, 113
Nitro-acetate of chrome for extract styles, 129
Nitro-benzol, 160

O.

OLIVE FOR INDIGO, 110
    (light) for indigo, 110
    for pigment colours, 126
    dye for wool, 58
Opal blues, 168
Orange dye for cotton, 54
    for indigo discharge style, 109
Orceine, 136
Orcine, 136

INDEX.

Orellin, 134
Organic matter in water, how to remove, 42
Oxalate and chromate standard, 109

P.

PADDING MACHINE, 39, 40
    purple (calico-printing), 99
    style, 89, 107
Paris blue, 171
Pastel, 159
    vat, 144
Peachwood, 157
Peacock blue for silk, 55
Pearl for steam styles, 120
Peonine, 189
Perkin's di-chlor process, 188
    method of preparing artificial alizarin, 187
    method of purifying anthracen, 179
    green, 175
Perrotine machine, 63
Persian berries, 157
    varieties of, 157
Persio, 140
Persoz's theory, 38
Phœnicienne, 189
Picric acid, 189
    red, 189
Pigment colour style, 93
Pink (light) for cotton, 53
    (bright rose), for cotton, 53
    (safflower), for cotton, 53
    for extract styles, 132
    for spirit styles, 122
    (cochineal) standard, for steam styles, 113
    cochineal medium, for steam styles, 113
    cochineal, pale or rose, for steam styles, 113, 120
    (magenta-aniline), dark, for steam styles, 113
    medium, 114
    pale or rose, 114

Pinkey's patent black, 45
Plaquage style, 89
Potassium, isopurpurate of, 189
Price's method of obtaining aniline violet, 177
Printed goods, how dyed, 74
    paper, bleaching, 29
Prussian blue dye, 49
Prussiate of tin pulp for steam styles, 115
Purple assistant liquor, 95
    (dark), for extract styles, 128
    (pale), for extract styles, 132
    fixing liquor, 95
    regina, 176
    Spiller's, 177
    for spirit styles, 122
Purpurate of ammonia, 156
Purree, 157

Q.

QUERCETIN, 141, 158
Quercitrin, 158
Quercitron bark, 158

R.

RAMSAY'S METHOD OF BLEACHING, 32
Red discharge on bronze, 124
    dye for cotton, 53
    for extract styles, 128
    for indigo, 110
    liquors, 153, 154
    dark, for machine work, 99
    pale, for machine work, 100
    resist, dark, for machine work, 100
    picric, 189
    for steam styles, 112, 117, 120
    brown for steam styles, 118
Regina purple, 176
Renard's method of preparing anilne red, 174
Reserves, 101
Reserve style, 88

INDEX.

Resist pastes, 89
   red liquor, 100
   style, 88
Rhein (chrysophanic acid), 158
Rhubarb, 158
Rose for extract styles, 132
Roseine, 172
Rosolic acid, 189
Rubbing machine, 18
   process, 18
Rubian, 153

S.

St. Martha's Wood, 157
Safflower, 158
   pink for cotton, 53
Safranine, 171
Saffron, 159
   adulteration of, 159
Safrosin, 191
Sage, pale (calico-printing), 105
Salmon for indigo, 111
   dye for wool, 59
Sandal wood, 158
Santalin, 158
Sapan standard for steam styles, 117
   wood, 139
Saxony blue, 146
Scald, the, 20
Scarlet dyes for wool, 59
Scotch plan of bleaching linen, 14, 15, 16
Sour, the, 20
Schuller's method of purifying anthracen, 178
Silk, bleaching, 27
   blue dyes for, 55
   to dye, 54, 55
   cleansing, 68
   forms in which dyed, 35
Silver drab (medium) for steam styles, 120
   (pale) for steam styles, 120
   standard for steam styles, 119

Singeing machine (Tulpin's), 68
Slate dye for cotton, 50
   (medium) for steam styles, 119
   (pale) for steam styles, 119
   for pigment colours, 126
   standard for steam styles, 119
Smith's method of preparing aniline red, 174
   of preparing aniline violet, 177
Soda hypochlorite (chloride of soda) 20
Solferino, 172
Soluble blues, 168
Spiller's purple, 177
Spirit colour style, 92
Standard red liquor, 100
Steam-colour style, prepared, 91, 111
   chest, 92
   style, unprepared, 92
Steeping process, 19, 20
Stone for steam styles, 120
Straw, bleaching, 30
Styles of calico-printing, 86
Substantive colours, 36
Sulphindigotic acid, 146
Sulphindylic acid, 146
Sulpho-muriate of tin, 112
Sulphuring woollen goods, 26, 27

T.

Tan for Pigment Colours, 126
Tessié du Mathey's method of bleaching, 32
Theory of bleaching, 31
Thickeners, 84
   list of, 82
   remarks on, 82, 83, 84
Thom's apparatus, 32
Thread bleaching, 21
Tin, sulpho-muriate of, 112
Tulpin's singeing machine, 68
Turkey red with discharges style, 90
Turmeric dye for cotton, 54
Tyraline, 172

V.

Vat, indigo, cold, 144
  indigo, potash, 144
  indigo, Schützenberger & De Lalande's, 144
  pastel, 144
  woad, 144
Violet benzyl, 177
  dye for cotton, 50
  imperial, 176
  naphthyl, 177
Violets, Hoffman's, 176

W.

White Discharge on Bronze, 123
Wash mills, 17
Water, choice of for dyeing, 41
  organic matter in, how to remove, 42
Wax, bleaching, 30, 31
Weld, 159
Woad, 159
  vat, 144
Wool to dye, 54, 55

Wool, forms in which dyed, 35
  black dyes for, 55, 56
  brown dyes for, 57
  chocolate dye for, 57
  crabbing, 68
  claret dye for, 57
  crimson dye for, 59
  drab dye for, 57
  green dyes for, 58
  salmon dye for, 59
  scarlet dyes for, 59
  olive dye for, 58
  yellow dyes for, 60
Woollen goods, bleaching, 26, 27

Y.

Yarns, Bleaching of, 12
  Irish plan of bleaching, 24, 25, 26
Yellow aniline, 177
  berries, 157
  discharge on bronze, 124
  dye for cotton, 54
Yellowish eosin, 191
Yellow, French, 190
  for indigo discharge style, 109
  for steam styles, 121
  Manchester, 190
  dyes for wool, 60

THE END.